The Living World

The Living World

Michael Bright

St. Martin's Press
New York

For information, address St. Martin's Press, 175 Fifth
Avenue, New York, N.Y. 10010

Library of Congress Catalogue Card Number: 88 43584

Design by David Dick

ISBN 0-312-03022-3

This book was first published in the United Kingdom by
Robson Books Ltd.

First U.S. Edition

10 9 8 7 6 5 4 3 2 1

Contents

Introduction

'THE LIVING WORLD' – BBC Radio 4's weekly exploration of the natural world – has been running for over twenty years. During that time, listeners have been able to visit many extraordinary locations, both familiar and exotic, and through the medium of words and sounds, uncover some of the amazing secrets of plants and animals. In the company of presenters Tony Soper, Derek Jones and Peter France, and under the guiding hand of producers John Sparks, Dilys Breese, John Harrison, Brian Leith, Ann Blair Gould, Melinda Barker and myself, the programme has kept listeners in touch with the latest wildlife news, both from the field and from the laboratory.

In this book, I have gathered together some of the wildlife stories that have especially interested me. It is by no means a comprehensive survey of recent developments in the natural sciences, but a pot-pourri of observations, discoveries and mysteries; and, there is a clear bias towards things animal, and things marine, for which I offer no excuses. As an ex-Plymothian, with the English Channel and the Atlantic Ocean on my doorstep and the fascinating aquarium of the Marine Biological Association virtually in my backyard (my nose was pressed firmly to the glass most weekends and school holidays), I have always found the sea, and the creatures living in it, particularly the marine mammals – whales, dolphins, seals – exciting, mysterious and awe-inspiring.

When writing down the stories, I noticed several themes emerging. There are the superlatives – the deepest divers and the fastest weaners – and the associations – ants and butterflies, honeyguides and badgers, and deer and monkeys. There are behaviour patterns and physiological adaptations that seem to have parallel developments in totally unrelated species: courtship behaviour at leks in humpback whales, hammer-headed bats, kakapos, and bower birds; heat-exchangers for body-heat retention in emperor penguins, great white sharks, swordfish and tuna; synchronous egg-laying in populations of horseshoe crabs, sea turtles and coral polyps; and migrations – polar bears, blue sharks, and caribou. There are animals doing unexpected things – elephants drinking alcohol, chimps taking medicine, bees making polyester, and piranhas eating nuts. Then there are the mysteries – the giant octopus, a living dinosaur, and the penguin graveyards.

Where appropriate, the researchers responsible for the new dis-coveries are acknowledged, but there are many, too numerous to mention lest the book become an endless list of names, who have contributed to our understanding of the lives of the animals mentioned, and whom I can only thank en masse for exciting me and the many listeners to Bristol's Natural History Unit programmes. Without their dedication and discoveries I would have nothing to write about.

1
The Arctic

WHEN THE GREAT NORTHERN forests are left behind, and the mean summer temperature does not rise above 10°C (50°F), you are in the Arctic. It is a cold and desolate place, with a short frantic summer and a long, dark winter. The shallow, frozen Arctic Ocean, covered for most of the year by 4m (13ft) thick pack-ice, is surrounded by frozen continents and islands – the northern parts of Scandinavia, North America, Siberia, Iceland, the whole of Greenland, Baffin Island, and Svalbard. Few animals live there permanently, but wave upon wave stream in at the time of summer plenty, when an explosion of flowers and insects on land, and a sudden bloom of plankton and small animals in the sea, provide food for the new generation. Waiting for the youngsters to appear, and ever ready to snatch an easy meal, are the predators. On land, timber wolves and Arctic foxes harass caribou and nesting birds while, in the sea, the killer whales and polar bears reign supreme.

The Ice Bear

Every spring and autumn, the townsfolk of Churchill, on the western shore of Hudson Bay, must not venture on to the streets at night. Their schoolchildren are ferried about in buses, and armed mounties and wildlife officers patrol the town during the day. The inhabitants, though, would not want it any other way, for Churchill is the polar bear capital of the world and the tourist revenue is worth over two million Canadian dollars a year. Locals wear sweatshirts that proclaim, 'Our Household Pests are Polar Bears'.

The polar bears pass through Churchill on their annual migrations, heading south in the spring to give birth to their pups, and moving north in the autumn to hunt seals in the Bay. A large rubbish dump on the outskirts of town is a convenient stopping-off point and it acts as a magnet to the migrating bears. Visitors from around the world have also converged on the dump to see one of the world's most exciting wildlife spectacles. Unfortunately, the bears do not always stay out at the dump but head uptown in search of more food. This was particularly

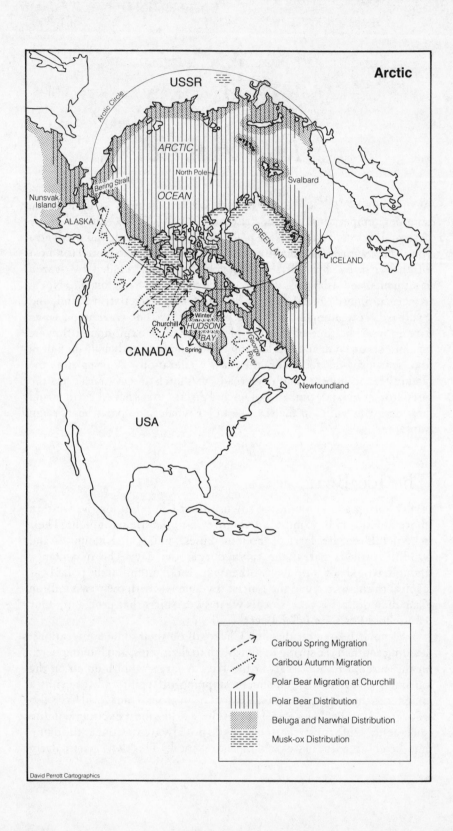

Arctic

USSR

Arctic Circle

ARCTIC

North Pole

OCEAN

Svalbard

Bering Strait

Nunsvak
Island

ALASKA

GREENLAND

ICELAND

Winter
Churchill HUDSON
BAY

CANADA

Spring

George
River

Newfoundland

USA

David Perrott Cartographics

⤏	Caribou Spring Migration
⋯⋯▸	Caribou Autumn Migration
⟶	Polar Bear Migration at Churchill
‖‖‖	Polar Bear Distribution
▦	Beluga and Narwhal Distribution
▤	Musk-ox Distribution

disturbing in 1983 when an unexpected spell of warm weather in the late autumn meant that the ice had not formed on Hudson Bay and the bears could not go hunting. Many hungry bears roamed Churchill and a middle-aged man was killed on the main street. Now, conservation authorities are tranquillizing the bears as they reach the dump and flying them off to the wilderness. The loss of bears has meant a decline in the number of tourists and the business community is up in arms.

Polar bears are not animals to fool about with. A male can weigh up to 650kg (1,430lb) and be 3m (9ft 9in.) long. One blow from its huge and powerful forepaw can cave-in the skull of a seal.

The polar bear is considered by some to be the symbol of the Arctic, but until quite recently its way of life has been a complete mystery. It was thought at one time that bears sit on ice-floes and drift clockwise around the North Pole, obligatory boreal nomads, at the mercy of the currents and eddies of drifting ice, eking out an existence from their sterile environment. Long-range tracking experiments, in which bears have been fitted with radio transmitters and their movements followed by satellites in space, have shown that this picture of polar bear life is quite wrong.

Polar bears do not, it seems, travel willy-nilly across the ice, but instead live in large familiar areas where they may travel as much as 40 km (25 miles) a day for many days. Alaskan bears, slightly larger than the rest of the Arctic's polar bears, move even further each day. Bears can also travel against the direction of ice drift. About three-quarters of the ice in the polar basin flows out between Svalbard and Greenland into the Greenland Sea, sweeping down the east coast of Greenland at a drift rate of about 25 km (15 miles) a day. The bears travelling on this moving ice can walk against the drift and navigate accurately back to their home bases.

They are solitary animals, only coming together to breed, although occasionally twenty or thirty animals will tolerate each other's close proximity at a particularly good food supply, such as a whale carcass.

Ringed seals are the preferred prey, and, if they are not available, bears will settle for bearded, harp and hooded seals. They have been known to attack a walrus or beluga, but more usually feast on stranded carcasses. Small mammals, birds and birds' eggs and even tundra plants have helped the occasional bear through hard times.

In different areas, bears have different preferences. Svarlbard bears eat mostly seals, whereas the Hudson Bay bears have taken up a lifestyle reminiscent of brown bears, with dens in the forest and a summer diet of berries and birds. This is interesting because it marks a regression to an ancient way of life. It is thought that the polar bear and the brown bear had a common ancestor, the Pleistocene bear. It was as recent as 20,000 years ago that the two lines diverged, and the predominantly carnivorous diet is thought to be a modern adaptation to life in the Arctic.

Seal hunting is carried out in a variety of ways. In April and May, for instance, bears might take advantage of the plentiful supply of ringed seal pups by breaking into their dens below the snow. Some Canadian bears have developed a novel way of hunting adult seals by swimming under them. The Svalbard bears are traditionalists. They wait beside an air-hole or creep up on a victim and swat it dead with one blow of the forepaw.

Mating takes place from April to June, with males locating receptive females by scent. Implantation of the fertilized egg is delayed and the cubs – a single, twins or triplets – are born in the female's winter den in January. The mother bear's milk has a high fat content which allows the cubs to keep their body temperature up during the cold winter and early spring, and rapidly gain weight during the four months the family remains in the den. Cubs stay with their mother for a little over two years.

The polar bear is recognized as *vulnerable* in the Red Data Book. This means that the International Union for the Conservation of Nature and Natural Resources (IUCN) have considered the species to be more than *rare* and moving towards the next category *endangered*. The next stage is extinction. Disturbance and habitat destruction from oil and mineral exploration is one serious threat to polar bears.

There is still a small aboriginal hunt allowed but even that is restricted because large concentrations of heavy metals, such as mercury and lead, and pesticides, such as DDT and dieldrin, have been found in the flesh making the meat unfit for human consumption. As yet, the bears themselves are not showing any signs of changes in behaviour caused by the pollutants. The chemicals, it is thought, are carried by the wind thousands of miles from the industrial complexes of North America.

Lords of the High Arctic

On 27 September 1984, a Canadian biologist was camped about 65km (40 miles) below the Limestone Falls on the Caniapiscau River in northern Quebec when he spotted a caribou carcass floating downstream. He thought no more about it, until the following day he saw another dead body, then another and yet another. He set out to investigate and the further upstream he went, the more dead and bloated caribou he found. Eventually, he came to a cove – since renamed Death Cove – just 16 km (10 miles) below the falls and before him 2,400 carcasses were piled one on top of the other. He realized immediately that there had been a major ecological catastrophe involving what was once one of the largest herds of caribou in the world – the George River herd.

The river flows northwards into Ungava Bay in the northeast corner of Quebec, and marks the boundary of the 'barren lands' where the herd spends the summer months. Come the first snows, when the topsoil freezes and feeding is difficult, the entire herd, 30,000 strong, migrate southeast to their winter refuge alongside Hudson Bay. In a long winding column they travel maybe 30km (20 miles) to 65km (40 miles) a day, their broad hooves giving a firm grip on the moss or snow. They are fat and fit after a summer of plenty, feeding on lichens (50 per cent of their diet) and the leaves, but not stems or woody parts, of low-lying tundra vegetation.

The route of the autumn trek takes the herd across two major rivers, the George and the Caniapiscau, and each year they use the same fording places: 1984 was no exception, except that the Caniapiscau was in flood and the normally dangerous Limestone Falls were impassable. Nevertheless, driven on by an instinctive urge to go southwest in order to survive the bitter winter months, the first few animals plunged into the icy waters.

In most years, maybe a hundred animals are swept away at the hazardous narrows where rocks, chutes and rapids funnel the water through the Falls. On 25 September 1984 and for the next three days 10,000 caribou were drowned or smashed against boulders.

Caribou are normally good swimmers – they are naturally buoyant and can propel themselves through the water, with their splayed hooves, at a speed faster than a human can walk. The George River herd, however, was decimated and those that managed to struggle to the other side of the ford pressed on to the Bay, unaware of the misfortune that befell the rest of the herd.

In the forest the snow is deep, but, because the sun does not penetrate the foliage to melt it and the frosts are less severe, it is soft and powdery. Using their front hooves, the caribou dig 'feeding craters' and nibble away at the lichens and sedges below the snow. The name 'caribou' comes from the Micmac Indian word *xalibu*, meaning 'the burrower'. In conditions of very deep snow, the caribou eat the lichens growing on trees.

The females and youngsters tend to stay in the shallower snow at the forest edge while the large males venture within. The caribou bodies are protected by thick fur that can withstand temperatures dropping to 45°C below freezing. There is also a unique physiological mechanism that prevents body cells from freezing-up.

The membrane that surrounds animal cells consists of a double layer of fatty acid molecules. The fatty 'head' ends of the molecules line up side by side, whilst the 'tails' wave freely, and, as long as the 'tails' are free, vital chemicals can enter or leave the cell. Under conditions below freezing, the tails link up and prevent this movement, and so starve the cell of nutrients. To overcome the problem, the caribou obtains a fatty acid called arachadonic acid, which has a tail with kinks in it, and substitutes this substance in its cell walls. The new fatty acid does not link up in extreme cold conditions, and so chemicals can flow to and fro as normal. This 'anti-freeze' is to be found in the one item that makes up over half the caribou's diet – reindeer 'moss', which is really a grey lichen.

In spring, between February and April, the herd moves out of the forest and starts the 1,000km (620 mile) return journey, through blizzards and frozen rivers, to the tundra. The herd follows the same route that it has travelled for centuries and in places the rocks may be worn down to a depth of 60cm (24in.) by the thousands of caribou hooves that have passed over them on countless migrations. Many of the cows are pregnant and they head the migrating column, sometimes strung out 300km (185 miles) behind the leading animal. Males and immature animals follow at the rear. It may take several weeks for the entire herd to cross a river.

On reaching the inhospitable tundra, the cows drop their calves. Here they are relatively safe from large predators, such as wolves, bears, lynx and golden eagles, which remain further south. Not long after calving the youngsters are able to run and feed. They must put on weight rapidly because the greatest threat to their early life is preparing to emerge from the millions of ponds and lakes on the surrounding land. At the beginning of the summer the tundra is plagued by myriads of mosquitoes, black flies, warble flies and other biting insects. They can take a litre of blood a day from a defenceless animal. Weaker animals may even succumb to the onslaught.

In July, the calves are weaned and the herd disperses to feed on the summer's new plant growth. In August the insects have gone and the herd fattens-up for the winter. In the autumn, the rut begins, with antler size being a significant factor in the determination of dominance. A bull who breaks an antler immediately loses his place in the hierarchy. The dominant males attempt to herd the females into harem groups. Gestation is about 220 days and a female usually has one calf which is born the following spring.

Unlike other deer, both male and female caribou have antlers. The males lose them in December, while the females retain them until May. Another curiosity is the 'twanging' or 'clicking' sound that caribou make when they walk. This is caused by the movement of a small tendon in the foot.

The caribou of the North American Arctic are still numerous and widely distributed across Canada and Alaska. They are known as the barren-ground caribou, and by far the majority are to be found to the northwest of Hudson Bay. There are, however, signs that even these animals, which constitute the most numerous herd-animals on the move annually, are being disturbed. The exploration and exploitation of oil and mineral resources have already interfered with the migration routes of some of the herds. The construction of the Alaskan pipeline, for example, despite the provision of crossing places, has seriously upset animals moving through the area.

The catastrophe that befell the George River herd was a local disaster and the result of an unusual series of events. Abnormally heavy rains in September 1984 resulted in the need to release water from a hydro-electric dam upstream from Limestone Falls. But it does demonstrate how vulnerable even the largest concentrations of animals are to our activities, and how a simple accident or careless act can be devastating.

The Bearded Ones

When danger threatens, a herd of musk oxen will form a defensive phalanx, known as the 'karre'. The largest cows and bulls close ranks to the outside while the youngsters are trampled and squashed on the inside. The herd shuffles and reels as one, presenting predators, such as a hungry pack of wolves, with an impenetrable wall of hooves and horns. Against a hunter's gun, of course, the defence is useless. Fifty years ago the musk ox was facing extinction from overhunting. In Russia and Alaska they were totally annihilated. Strict conservation measures, though, have ensured their continued survival, with herds introduced successfully to four Alaskan islands, Scandinavia, and Russia. Game reserves, such as the Thelon Game Sanctuary in the Canadian Arctic, have afforded some measure of protection against hunters and the herds have been able to breed and increase their numbers.

The musk ox is a huge and hairy creature. It has two thick coats to keep out the icy cold of the Arctic winter – an inner one of dense wool known as 'qiviut' which is said to be softer than cashmere, and an overcoat of coarse guard hairs that hang down in a long, shaggy skirt. The Inuits (Eskimoes) have given it the name 'Oomingmak', meaning 'the bearded-one'.

The horns, heavy and downswept like meat-hooks, are larger in the bull. The animal looks like, and indeed is, a survivor from a time when ice sheets advanced and retreated across the whole of the northern hemisphere, and woolly mammoths vied with sabre-toothed cats for dominance on the desolate plains of Siberia, Northern Europe and North America.

Curiously, in late spring some of the musk oxen of Nunivak Island, off the west coast of Alaska, attempt to leave dry land and cross the melting ice. They are, perhaps, seeking new pastures and relieving over-crowding on the island. Many perish, for, when the ice breaks up, they become stranded on ice-floes and starve or freeze to death.

The 'musk' in musk ox refers to the musk gland of the male and is much in evidence during the rut when the males smell strongly and are very aggressive, settling disputes about dominance by bouts of head butting. The name 'ox' is inappropriate for the musk ox is one of the goat-antelopes related more closely to the takin of the Himalayas. Unlike the North American caribou and the reindeer of Europe, the musk ox does not migrate to the south during the harsh winter months. Instead it braves the cold and biting winds, protected by its thick fringed rug. When a blizzard is blowing its strongest, a herd will form into a wedge shape, bulls at the tip and cows behind with the calves protected in the centre.

In summer the musk ox has an overheating problem and will often find patches of snow in which to roll to keep cool. Unseasonal high temperatures can be disastrous. Fog or rain followed by re-freezing can mean ice crystals form in the fur, and if an animal cannot shake them out it becomes so heavy that it cannot move and becomes vulnerable to wolf attack.

Momentary Motherhood

The hooded seal is the largest of the true seals living in Arctic regions. It spends much of its time at sea, only coming out on to ice-floes to pup. Its name is derived from the black bulbous structure on its nose. This it inflates to impress and deter aggressors. It can also blow the nasal septum out through the left nostril to form a large red bladder.

The most remarkable thing about hooded seals, however, is the speed with which the young are raised. From birth to weaning takes just *four* days. This is the shortest period known for any mammal. It has been known for some time that seals pupping on the Arctic pack-ice rear their pups very rapidly. The harp seal, for example, takes about ten days and other seals about two to three weeks.

Hooded seal mothers produce very rich milk, with a high level of butterfat, so that, in the four days, the pup is able to double its body weight from 20 to about 40 kg (9–18 lb). It is born at a fairly advanced stage and very late in the pupping season. Of all the Arctic seals, the female hooded seal is the last to drop her pup. Usually the ice is beginning to break up and so she must raise her offspring quickly, so that it has sufficient insulating blubber to allow it to enter the water safely before the ice-floe, on which they are both sitting, melts and disappears.

The reason for the late pupping and fast weaning is not clear, although it might be that there is an abundance of food earlier in the season on which the female can fatten up in preparation for rearing her young. A short and intense period of suckling would also mean that the female loses less body weight, and she would not need to leave the pup in order to feed. In addition, a late birth reduces the likelihood that predators, such as polar bears, would cross the unstable pack-ice and take the young.

The Unicorn of the Seas

What looks like a pair of submerged swordsmen duelling for the hand of a maiden is, in reality, two mottled grey male narwhals fighting for dominance and the right to mate with a herd of females. Male narwhals have been seen to 'cross-swords' as a preliminary to courtship and mating. The fights can be bloody – most large males have scars about the head to prove it and many have broken tusks.

The sword itself is curious. It is offset to the left and points downwards. It can grow up to 3m (10ft) in length which makes it a rather awkward protuberance to be carried about by a whale only 5m (15ft) long. The tusk is an elongated left canine tooth that grows in a tight spiral out through the upper lip. The narwhal's only other tooth remains small, although in some individuals both teeth grow into long tusks. In the female one of the pair of functionless teeth grows into a small 20cm (8in.) tusk.

The narwhal was brought to the attention of the scientific community after the return of Martin Frobisher from his search for a nothern Atlantic–Pacific sea route. In July 1577, his three ships were sheltering from a storm in what is now known as Frobisher Bay, when crew members discovered 'a great dead fish' that was 'round like a porpoise, being twelve feet long' and it possessed 'a horn of two yards length'. Frobisher thought they had found a 'Sea Unicorn' and the long piece of spiralled ivory was presented to Queen Elizabeth I. What Frobisher had found, of course, was not a fish but an air-breathing marine mammal and a relative of the beluga or white whale.

Magical medicinal properties were attributed to the narwhal's elongated tooth. Crushed tusk was said to 'expel ill vapours by sweate'. It was also recommended for strengthening the heart and curing epilepsy. Indeed it was prescribed by leading physicians up until the late eighteenth century, providing a healthy business for Arctic whalers who were able to sell narwhal tusks for their weight in gold. The seventeenth-century Danish anointment throne in Rosenbary Castle, Copenhagen, features narwhal tusks and, according to a Danish bishop, it is more splendid than Solomon's ivory and golden throne.

Today the narwhal is protected, although a small catch of 542 animals by Inuit hunters is allowed by the Canadian Government. Trade in horns was reduced by the US Marine Mammal Protection Act of 1972. Until the European Economic Community banned the import of narwhal tusks in 1983 a good specimen could raise as much as $1,000, but only clients dealing with dedicated collectors will secure a high price. In New York a tusk was reported to have changed hands for $4,500. Greenlanders still eat the meat and the people of Baffin Island relish the skin, called 'muktuk', which is rich in vitamin C.

Little is known about narwhal biology, migration or distribution and population sizes. They are usually seen in groups of about ten to twenty animals moving at the edge of the pack-ice, although mariners have reported seeing thousands of animals on the move. Robert Peary, for example, witnessed narwhals 'dashing to windward, their long white horns flashing out of the water in regular cadence'. In winter they move out into deeper water and to the south. In spring, when the ice breaks up, they migrate inshore into fjords and estuaries, often alongside white whales. They eat cod, halibut, flounders, squid and shrimps which they must suck into their mouth and crush between the jaws because their two teeth have no grasping or chewing function. They make whistling and clicking sounds while travelling. The whistles are thought to be communication sounds between individuals in a herd and the clicks are likely to be echo-location signals used for pinpointing food.

It is estimated that at present there are over 30,000 narwhals, divided into three distinct populations, living in Arctic seas. The largest group is 20,000 strong and lives in the Baffin Bay–Davis Strait area. Although they are very much polar sea inhabitants they have been found occasionally further to the south. One was even found stranded 30 miles up the estuary of the River Thames.

The Sea Canary

Belugas or white whales might be considered honorary song birds. They are certainly the most vocal of the small whales with a large repertoire of whistles, grunts, and clicks that have earned them the nickname 'sea canaries'. William Edward Parry, the British polar explorer, wrote in

1819 about his arrival at Lancaster Sound where bowhead whales, walruses and narwhals were seen with white whales 'swimming about the ship in great numbers'. He described 'a shrill, ringing sound not unlike musical glasses played badly'. Other observers through the years have variously described white whale sounds as the mooing of cows, the gritting of teeth, human screams, rusty hinges, the ringing of bells, and the crying of babies. Most of the sounds are thought to be used for communication. Whalers have noticed that when one whale is harpooned all the other white whales in the area, even including those in the next bay, disappear from the surface. The sounds must travel considerable distances and be picked up by whales miles away from the caller. A clapping of the jaw is a particularly loud sound peculiar to belugas. The clicks are thought to be echo-location signals.

The white whale is unusual amongst whales in being able to turn its head to the side and also change the shape of its mouth. It is thought to purse its mouth into an 'o' shape when sucking up food from the bottom of the sea. Food can be almost anything. A whale's daily requirement of fifty to a hundred pounds of food may be made up from bottom-dwelling worms, shellfish, sculpin, or flatfish and midwater salmon and Arctic cod.

In winter the white whales are to be found in polynyas – areas of open water, free from ice – and in the spring thaw they migrate inshore, sometimes in herds numbering thousands, to the bay, estuary or inlet where they were born. They come in July to give birth, to moult and to play. Here the water is a little warmer for the more sensitive new-born calves. It is, perhaps, significant that the whales do not congregate at the mouths of glacier-fed rivers where the sea is far colder. An adult whale is insulated with a 15–20cm (6–8in.) thick layer of blubber whereas a youngster has barely an inch. The estuaries are also comparatively safe from their main predator (apart from man) – the killer whale. White whales swim at a top speed of about 15kph (9mph) and are unlikely to outrun the highly manouevrable orca that can rush into the attack at 48kph (30mph). The shallow, tidal waters are white whale havens.

After fourteen months in a warm, safe womb the 1.5m (5ft) baby beluga emerges head first into water at a temperature of 10–16°C (50–60°F). It is nudged to the surface by its mother or a helping 'aunt' in order for it to take its first vital breath. From then on it travels close to its mother, sometimes lying across her back or close to her head. It puts on fat rapidly, taking milk that is 35 per cent butterfat.

Before their return to the open sea, the adult belugas appear to moult. Inuit hunters tell of white whales entering estuaries in the spring with yellowing skin and leaving again in the autumn pure white. It is thought that the whales swim into gravel banks on the rising tide and rub the old skin off on the stones. This can be dangerous, for if they mistakenly roll in the shallows on a receding tide they can become stranded. At this time they also indulge in another curious piece of behaviour.

Males have been seen to swim to the bottom, pick up a strand of seaweed in the mouth, and rush about with it like a streamer while all the other males are in hot pursuit. During this game of tag, the males attempt to tear the frond to shreds. Sometimes a male will grasp a stone in its mouth or balance it on its head. Again, all the others bump and chase as if trying to dislodge the stone. It has been suggested that this playful behaviour is simply a 'game' being played by intelligent animals.

Females have been seen with small planks of driftwood on their head or back. This strange behaviour is thought to stem from a mother's habit of supporting a new-born baby or a youngster that is in trouble. A female that has lost her calf may adopt a piece of wood as a baby substitute.

By mid-August the young whales will have doubled the thickness of their insulating layer of blubber and the adults will have completed their moult, and so all the whales once more return to the open sea. Some get lost. In 1932 a white whale reached the Forth Railway Bridge in the Firth of Forth before it was killed, and in 1966 another swam up the Rhine as far as Bonn, spending a month in the busy and polluted river, before it returned safely to the North Sea. In 1980 a dead beluga was washed ashore at Formentera, one of the Balearic islands of the Mediterranean.

Today, about 100,000 white whales live in Arctic seas. Their numbers are still recovering slowly from the slaughter at the turn of the century when the larger filter-feeding baleen whales, particularly the bowheads and the right whales, had been hunted almost to the point of extinction and the hunters directed their harpoons at the slow-swimming white whales. Driven into shallow bays, they were killed in their thousands. The fat from six 5.5m (18ft) belugas gave about one tonne of oil. The skin, made into belts and boot laces, sold in Britain for one shilling and sixpence a pound. In the early days of whaling, white whale oil was used to top-up the oil from the less easy to catch baleen whales. Currently, subsistence hunting takes place in Russian and Canadian waters, although there have been restrictions placed on any commercial exploitation because high levels of mercury have been found in the flesh. White whales are more likely to meet man-made pollutants because of their habit of spending time in coastal waters.

2

Antarctica and the Southern Ocean

There is no major land mass within 3,220km (2,000 miles) of the Antarctic continent. It is surrounded by pack ice and the icy clockwise swirl of the Southern Ocean. All of the Antarctic animals, and those living on sub-Antarctic islands, are dependent upon the sea, and only a few species, mainly seabirds, visit the mainland, during the summer months, in order to breed.

The continent itself is a system of high plateaux ranging between 1,980m (6,500ft) and 3,960m (13,000ft) and mountain ranges, with occasional peaks reaching 5,180m (17,000ft). The surface is covered by nine-tenths of the earth's ice and snow, and the slow run of fresh melt-water influences physical conditions in the surrounding sea. Cold, brackish water from the south meets the warmer, more saline, water from the north, and where they meet is the Polar Front. Here, the warm water rises over the cold bringing nutrients to the surface. To the north of the 40km (25 mile) wide front, approximately latitude 50° South, life is sparse, whilst to the south there is probably the greatest concentration of animal life on earth.

Land of Extremes

Curiously, some parts of the Antarctic are the driest places on earth. With less than five millimetres of precipitation, all falling as snow, the 'dry valleys' are estimated to be 15 times more arid than the middle of the Sahara desert. They are also cold, very cold, and in the winter months there is perpetual darkness.

Temperatures in the dry valleys of Victoria Land regularly fall to minus 60°C in winter, and in central Antarctica a minimum of minus 88.3°C has been recorded. In summer the temperature, for the most part, remains below freezing. Winter winds reach 300kph (186mph), pushing the fine, powdery snow into drifts that can leave large areas snow-free. Humidity is so low, that, come the summer thaw, the snow does not melt but misses out on the water phase and evaporates immediately – a process known as sublimation, and the principle of freeze-drying.

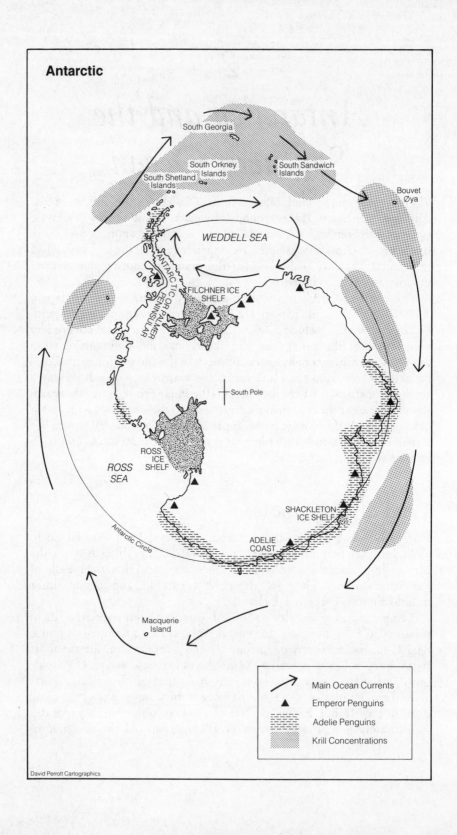

Antarctic

South Georgia

South Orkney
Islands

South Shetland
Islands

South Sandwich
Islands

Bouvet
Øya

WEDDELL SEA

ANTARCTIC OR PALMER PENINSULAR

FILCHNER ICE
SHELF

South Pole

ROSS
ICE
SHELF

ROSS
SEA

SHACKLETON
ICE SHELF

ADELIE
COAST

Antarctic Circle

Macquerie
Island

Main Ocean Currents

Emperor Penguins

Adelie Penguins

Krill Concentrations

It is also a salty environment, the consequence of chemical weathering of the local rocks and the lack of water to wash it away. It is the harshest and most extreme environment on this planet, and some have considered it to resemble conditions on the surface of Mars.

Yet, there is life here. On the surface and in the upper layers of the soil live algae, fungi and bacteria, each darkly coloured in order to absorb as much of the sun's radiation as possible. Blue-green algae combine with fungi to form pitch black lichens.

In the shallower parts of dry valley salt lakes, thick mats of blue-green algae cover the bottom. If freezing reaches the floor of the lake, algae become incorporated into the ice. In the spring, when the ice starts to melt, small air bubbles appear around the plates and miniature 'glasshouses' are created inside the ice, each with its own ecosystem. Here, nematode worms, microscopic protozoans, rotifers and bacteria and algae live in an enclosed environment that can have a maximum temperature of about 20°C (68°F).

In Lake Fryxell, at the eastern end of Taylor Valley in southern Victoria Land, stromatolites have been found. These are dome-shaped mounds of calcite overlain and deposited by a living mat of blue-green algae and diatoms. They resemble structures found as fossils in rocks from the Precambrian era, about 2,900 million years ago.

On the underside of quartz crystals, which lie scattered over the valley floor, mosses, lichens and algae grow upside-down. It is thought that the translucent crystal acts as a lens to focus the sun's rays on the plants growing below.

On the valley sides, there is a zone of lichens that grow in cracks and crannies in the rocks. There is also a community of plants, again mostly associations of algae, fungi and bacteria living in bands under the surfaces of light-coloured and translucent rocks, such as sandstones, marbles, quartzites and some granites. Light penetrates the outer layers to facilitate photosynthesis, and the niche is only occupied on north-facing slopes.

Some of the most peculiar finds in dry valleys are the mummified carcasses and bleached skeletons of seals and penguins. In the 1957-8 summer exploration season an American team discovered 90 mummified crab-eater seals in the Taylor Valley, and radio-carbon tests showed them to be 1,600-2,000 years old (the dates are a little suspect because the source of the ancient carbon could have been sediments deposited thousands of years ago and only recently released in an upwelling, when it has entered the food-chain and ended up in a seal). Why they should travel so far inland is not clear, although sick animals and those bulls injured in fights during the breeding season are known to travel up to 56km (35 miles) away from the sea, safe from marauding killer whales, to rest and recover.

Harvest from the South

The Antarctic or Southern Ocean is said to be the most productive ocean in the world. The Atlantic and the Pacific have their 'upwellings', such as the Benguela Current off southwest Africa and the Humboldt Current along the Pacific coast of South America, where nutrients from the sea bottom rise towards the surface providing food for plankton which is the first stage in the ocean food-chain, but the Southern Ocean, surrounding the continent of Antarctica, has a more widespread distribution of vertical mixing of nutrients and thus an enormous profusion of life. The main character in the Antarctic food-chain is krill.

Krill is a small shrimp-like marine crustacean, 5–6cm (2in) long when fully grown, which lives in huge concentrated shoals in the Southern Ocean. In other parts of the world, the name 'krill' is given to other marine creatures – minute pelagic red crabs in Chilean fjords, mysid shrimps off Vancouver Island, small fishes in the north Atlantic, and copepods in the Bay of Fundy – but in the south it refers solely to the euphausiid shrimp *Euphausia superba*. So numerous is krill, that it is the stable diet of the great baleen whales – that is, those remaining after the slaughter of the past eight decades.

Antarctic krill is a filter-feeder. It has the first six pairs of body appendages modified as a filter basket, and water, rich in drifting phytoplankton, is wafted into the structure at the leg tips to be strained at the leg bases. Recent research has revealed that the krill does not feed exclusively on floating plant material, such as diatoms or algae, but instead is an opportunistic omnivore, occasionally taking animal food, such as pteropods or sea butterflies – small marine snails with a foot modified as a wing – worms, other crustaceans, and even individuals of their own species.

The rear five pairs of legs are paddle-shaped and serve to push the creature through the water. It must swim continuously throughout its life or sink into the ocean depths where food is scarce.

Light appears to play an important role in krill life. The large black compound eyes are stalked and there are ten luminescent organs on the eyes, at the bases of one of the pairs of legs, and on the underside of the abdomen. These can be lit for several minutes or flashed on and off. It is thought that light signalling could be the mechanism by which krill can find each other in the vastness of the Southern Ocean and aggregate into their enormous shoals. It also might allow individuals to communicate across a shoal, perhaps, at times when animals want to get together for reproduction.

During the female's moult, when the exoskeleton is soft, the male transfers two ball-shaped packs of sperm to a cup-like depression on her underside. The sperm are extruded from the packs and are stored in a special chamber until the eggs are ready. The female krill carries her pink egg-bundle below her abdomen until it is time for spawning. This usually takes place close to the surface during the austral summer. As each egg passes a slit in the sperm storage chamber it is fertilized, and as many as 2,000 may be released in a couple of hours. From November onwards a female produces up to 20,000 eggs in several separate batches.

After all her eggs have been released the female does not die, as it was once thought, but she moults to rid herself of the sperm packets. Her ovaries, meanwhile, return to their earlier state and produce new eggs. It is not known how many times she might spawn in her lifetime.

The eggs slowly sink and the embryos gradually develop as they go progressively deeper. The miniature, oval hatchlings are known as 'nauplius larvae' and, when their yolk sacs are used up, at the fourth larval stage, they reverse the downward movement, returning to the surface waters where they are distributed in the surface currents.

From egg to adult takes about three to four years, with many moults and (depending which authority you consult) about five, twelve or fifteen identifiable stages. Distribution in the ocean is seasonal, both horizontally and vertically. When danger threatens a shoal of krill quite literally 'jump out of their skins'. All that a predator is left with are the discarded exoskeletons, feathery decoys floating in the water, while the shrimps have swum off rapidly in the opposite direction.

Adult krill live in enormous shoals, usually 40–60m (130–197ft) across but can be up to 600m (2,000ft), that stain the water red by day and light up with a blue-green fire at night. They can be spotted from ships, aircraft and even from satellites in space. With the demise of the whales, it was thought that there might be a glut of krill available for commercial exploitation, and several years ago five countries sent pilot industrial expeditions to the Southern Ocean to explore the potential. There was some considerable excitement at the time for krill can be located easily, hauled aboard in huge nets, and yield concentrated protein and vitamin A. In one report it was estimated that the world's fish catch could be doubled and cheap protein would be available to feed the starving.

In reality, krill has limited uses, primarily dried, ground-up and fed to livestock, although poultry fed on krill-meal lay red eggs. Unfortunately, the strong fishy flavour of dried or frozen krill is unacceptable to societies that are not used to eating marine fish and so it is not likely to be a significant food source for the poor and starving. In Japan it became a luxury food, although this traditional fish-eating nation considered the flavour of fresh krill to be so bland that krill was relegated to fattening up sea bream, red snapper and yellowtail at fish farms. Frozen krill has found another use as 'chum', or bait thrown into the water, for attracting large fish to sports-fishermen.

Concern was also expressed about whether it was sensible to exploit krill and several reports have questioned the existence of a 'surplus'. It is possible that removing krill commercially would be the last nail in the coffin of the blue whale – populations are showing few signs of recovery after many years of full protection. In addition, krill is the principle source of food for a great number of creatures, living both in and on the sea, in the southern hemisphere.

Any damage to krill stocks would have devastating effects. Blue, fin and humpback whales, crab-eating seals, seabirds, squid and fish feed on the krill; Adelie and emperor penguins and Weddell seals feed on the squid and fish; killer whales take baleen whales, crab-eater seals, and penguins; leopard seals prey upon seals and penguins; and so on in one great and complex interdependent food chain. Skim off the krill and the food chain collapses.

Another threat is mineral exploitation on the Antarctic mainland. If oil or any toxic chemical is spilt into the ocean, then krill is going to be the most vulnerable organism.

The Fish that Cares

If an animal makes a sacrifice, even the supreme sacrifice of life itself, to protect and maybe save the life of another, it is said to have shown altruistic behaviour. There have been many examples quoted in the past – alarm calls that warn others yet draw attention to the caller, or the shielding of young animals from predators with obvious dangers to the protector – but all have turned out to be acts that are not strictly altruistic for they prove to be beneficial to the animal involved, either in the protection of its own genes in relatives and offspring, or the expectation of reciprocal action from the animal being protected.

But, a young research student, diving in the icy waters off the Antarctic Peninsula, has found a fish that seems to show altriusm. It is a bottom-dwelling sculpin-like fish *Harpagifer bispinis* which lives at a depth of about 9m (30ft) and guards its eggs. The female fish lays the eggs during the austral summer and stays with them for five months, leaving only for short feeding excursions. Predators will quickly gobble them up if the fish is away for long, and fungal diseases can destroy the entire brood if left for more than a fortnight. The guarding fish is itself exposed to danger from larger fish and seals.

The surprise comes when the female is removed from the nest site. Instantly another male, unrelated to the female, will come in and hover

over the eggs. If that fish is removed, yet another male will do the same thing. The males are not kin to the offspring developing in the eggs, and so there is no parental or family interest, and the guardian is unlikely to get any reciprocal protection from the youngsters when they hatch. It might be that the males have lost their way, when returning to their own nest site, and simply guard any unprotected nest which they stumble upon. But, superficially at least, it seems that the guards have nothing to gain from their action – an example of altruism?

Deep Diving Seals

After a deep dive, a human diver returning rapidly to the surface, without decompression stops on the way, will almost certainly suffer from the 'bends'. The decrease in pressure results in the formation of nitrogen bubbles in the blood and body tissue, accompanied by considerable pain, particularly in the joints. If the diver does not undergo an immediate programme of decompression, by either sending him back down and bringing him up slowly, or by placing him in a decompression chamber, he will most probably die.

Many marine mammals – seals, sealions, dolphins and whales – dive to great depths and return rapidly to the surface. Why do they not get the 'bends'? A team of West German scientists, working in the Antarctic with Weddell seals, have found the answer.

They have developed a technique of sampling the seal's blood by attaching a microprocessor to its back. They found that the seal would breathe out before diving, and when it reached a depth of 30m (98ft) its lungs collapsed. Any residual nitrogen is pushed into the upper respiratory tubes and is therefore unable to pass into the blood. All the nitrogen already in solution in the blood is taken up by body tissues and there it stays until the seal returns to the surface, when the lungs expand and the nitrogen in the tissues returns to the blood.

Weddell seals are very deep divers, with recorded dives of 610m (2,000ft) lasting up to an hour. It is one of the largest species of true seals and has a large body with a remarkably small head. The males appear to vigorously defend underwater territories, usually strips up to 200m (650ft) long that follow the cracks between ice-floes. Their simplest tactic is to prevent rivals from using the gap as a breathing hole; these must then seek a break in the ice elsewhere. Below the gap, they perform their highly vocal display to impress the females on the surface.

The Weddell seal lives the furthest south of any seal, moving north in winter and south in summer with the edge of the main body of ice. It rarely comes on to land. Pupping, for example, takes place on stable ice.

Often adult teeth are worn down from cutting air-holes in the Antarctic sea ice. Their main enemies are killer whales which will sometimes try to tip them off the ice-floe on which they are resting.

Killer Seal

The leopard seal looks more like an ancient reptile, a leftover from the age of dinosaurs, than a sea mammal. It has a large head, set on a 3m (10ft) sleek body, and a huge mouth well-equipped with rows of vicious-looking teeth. Little is known of its life, for researchers only tend to see it feeding during the summer months.

It is almost always seen alone, swimming near the boundary of the Antarctic pack-ice during the summer and migrating as far north as Australia and South America in winter. Occasionally several leopard seals are seen on the same beach but they remain well spaced. The only large aggregation known is on the beaches of Heard Island, to the south of the Indian Ocean, where up to 700 appear during the winter.

The animal's reputation for ferocity comes from an insatiable appetite for just about anything that moves in the Southern Ocean. The leopard seal is most commonly found in the proximity of penguin colonies. The penguins can out-swim the seal in short bursts and escape back on to land where this well-adapted sea mammal is at a disadvantage. But if the penguin can be headed off and the pursuit extended, then the seal is likely to catch it. Grabbing the struggling bird in its jaws, the seal shakes it violently and, often as not, will swallow it, minus the feet and head, whole. The seal's gullet does not have a rounded cross-section, but instead is a flattened tube that can accommodate the smaller species of penguins. The larger king penguins are dealt with in a different way. The bird is shaken violently until the skin splits and then with one powerful flick of the head it is unwrapped, the skin peeling off either up to the chin or down to the legs. The seal eats the large breast muscles and discards the rest.

King penguins, having spotted a hunting leopard seal, will slap the water with their flippers and flap noisily back to the shore or ice-floe, thus warning their fellows and confusing the seal. They also trigger off these 'panics' for no apparent reason and a seal, although undetected, might simply be unlucky. Then, it adopts a different 'sneaky' strategy. It waits, just below the edge of the ice, for a penguin to enter the water and seizes the unsuspecting bird before it knows what has hit it. On other occasions it might leap straight out of the water, land in the midst of a group of penguins and grab the nearest one before returning to the sea.

Although penguins are the prey most often seen taken by the leopard seal, it also eats fish, squid, the occasional unwary seabird, and the young of other seals. It might also take chunks out of a whale or seal carcass, and has been known to eat the young of its own species. Young leopard seals have been seen to practise their predatory skills on the young of elephant seals.

Man is not thought to be of any gastronomic interest to the leopard seal, although it does have the unnerving habit of coming alongside and extending its neck to peer into a small boat. It might also chase along behind. They are simply curious.

Emperors of the Antarctic

In the not too distant past there were penguins over 1.5m (5ft) high, but they are extinct now, and the world's largest living penguin is the emperor which stands a little over a metre (3ft) and weighs 30kg (66lb). It is highly adapted for life in the sea, having wings modified as flippers to 'fly' through the sea, a streamlined body coated with a thick layer of fat overlain by short, fine feathers, and feet set well back which, together with the tail, act as a rudder.

The emperor is the deepest diving penguin so far recorded. It can stay below for up to 18 minutes and reach a depth of at least 265m (870ft). Its close relative, the slightly smaller king penguin, has been tracked down to 240m (787ft). This study was carried out by John Croxall, of the British Antarctic Survey, and colleagues at the Scripps Institution of Oceanography in San Diego. They were able to clamp tiny depth recorders to six adults which were about to go to sea to catch food for their young. The birds were out for periods between four and eight days, and half their dives were more than 50m (164ft).

These 13kg (29lb) birds require 2.5kg (6lb) of squid each day just for themselves; they must fish a further 3kg (7lb) each day for the youngsters. On each fishing trip, then, an adult must catch between 50 and 90 squid weighing 150–200g (5–7oz) each. With an average of 865 dives per trip, this means that king penguins catch squid on fewer than 10 per cent of their dives.

Emperor penguins are probably the only birds never to set foot on land, for even mating and egg-laying take place on the sea-ice close to the Antarctic continent. Unlike the other penguins they start their breeding in the autumn when the ice is forming. When the Adelie penguins, which frequent the same areas, move north, the emperors head south. They may travel considerable distances. Emperors breeding in the Bay of Whales in the Ross Sea have been found with recognizable pebbles in the stomach that could only have come from sites over 550km

(340 miles) away. And for some, reaching the ice edge is only the beginning of their journey, for rookeries can be as much as 160km (100 miles) from the sea. They can travel quite fast by lying on their bellies and tobogganing across the ice.

At the rookery, courtship is a noisy affair; the 'trumpeting' call can be heard up to a kilometre away. Birds pair-up but do not have their own territory to defend, simply 'personal space'. The female lays a large white egg, 12.7cm (5in.) long, which is immediately scooped up on the feet of the male. He covers it with a special fold of skin at the base of his abdomen, for it is the male who incubates the eggs. The female leaves her mate, and sets out on the journey back to the sea in order to stock up on food, mainly squid and fish.

By this time the perpetually dark winter has set in, and icy winds and blizzards, sometimes blowing up to 200kph (124mph), send the temperature down regularly to minus 50°C and even occasionally to minus 75°C. The males huddle together, each with its bill resting on the bird in front, in a great circular 'tortue' which slowly rotates as several thousand birds press to get closer to the centre. In the huddle the birds need use less energy to stay warm – a bird in the centre only uses up 100g (3½oz) of stored fat each day whereas a bird in the open would use 200g (7oz). When the blizzard subsides, the huddle disperses and the males space themselves out once more. If a male drops his egg, other males without eggs will rush in and scoop it up and start to incubate it. Such is the urge to have an egg, that occasional fights break out.

The male rides out the winter storms for two or three months and throughout the incubation he just shuffles around, cradling the egg on his feet inside the skin fold, and does not eat. Heat loss from the exposed part of his feet is kept to a minimum as the penguin has a clever heat exchange system in the legs that diverts heat away from the feet. The outgoing arteries are right next to the ingoing veins and heat passes from one to the other.

Eventually, the female returns fat and brimming with food to present to the newly hatched chick. If it hatches before her arrival, the male can summon up a milky secretion on which the 15cm (6in.) tall chick can feed for its first few days. The female then takes over the incubation, using the same method, and the male goes to sea to feed – he will have lost half his body weight during the winter fast. But he will return quickly to help feed the great, brown, fluffy chick. It gains weight slowly and it might take up to four months of feeding by both parents before the youngster is fledged and ready to take its place in the world.

There are over 30 colonies scattered around the Antarctic, ranging from 500 pairs on the ice at the Dion Islands off the Antarctic Peninsula to over 25,000 pairs near Coulman Island. How any survive at all in such

harsh conditions is a miracle. In 1901, when the first colony was found at
Cape Crozier, Edward Wilson of the British Antarctic Expedition
discovered that the mortality rate for one breeding season was 77 per
cent and he wrote that he considered the behaviour of the emperor
penguin to be 'eccentric to a degree rarely met with even in
ornithology'.

Penguin Graveyards

What happens to penguins if they avoid all the predators, survive the
appalling weather conditions and live to be a ripe old age – perhaps
twenty years for the large emperors and kings? There are a couple of
clues, albeit tentative.

In 1947, Neill Rankin noticed well-worn paths winding up on to
higher ground from a colony of king penguins on South Georgia. On
investigation, he found that they led to a small pool around which
several penguins were standing, all facing inwards, towards the centre.

Thirty four years previously, an American ornithologist, Robert
Cushman Murphy, wrote of his experience with two colonies of gentoo
penguins on the same island. He described finding a number of 'sick and
drooping penguins' standing around a pool of fresh melt-water, about
3m (10ft) deep. At its bottom he saw layer upon layer of gentoo bodies.

3

The Living World of Whales and Dolphins

Between 40 and 50 million years ago, some land mammals, living beside the ancient Tethys Sea, found that fishing was more productive than chasing after small land animals and returned to a life in the sea. Their remains were found in fossil beds in Pakistan, alongside crocodiles, catfish and other animals that would have lived in shallow estuaries. They were the ancestors of the whales and dolphins.

Today, we have two groups of cetaceans. There are the baleen whales (characterized by the two fringed, horny plates hanging from the roof of the mouth with which they strain the seawater for krill and small fishes), including, the grey whale, right whales, and rorquals – minke, Bryde's humpback, sei, fin – and the largest animal ever to have lived on earth, the blue whale (about the weight of 33 large bull elephants and the length of a couple of buses); and there are the toothed whales, including the dolphins (the largest of which is the killer whale), porpoises, white whales, beaked whales, and the enormous sperm whales.

They breathe air with lungs, as we do, and suckle their young, but, instead of hair, they have thick layers of blubber as insulation. They have streamlined bodies and are fully adapted to a life in the sea.

The Singing Whales

The children of Lir, so the legend goes, were turned into swans by their stepmother and banished to an island in the Atlantic Ocean. There they languished for 800 years, and, through the mists of time, seafarers and fishermen have heard their mysterious and eerie songs. Might that island have been Bermuda, and might those swans really have been whales?

In 1967, biologist Roger Payne, of the New York Zoological Society, was floating about on the ocean in a rowing boat off Bermuda, studying the migration of the great whales, when he heard the most exquisite sounds coming up through the bottom of the boat. He had heard those sounds before – on an underwater tape recording made by acoustics engineer Frank Watlington. They were the songs of the humpback whale.

Female Weddell seal sniffing her newborn pup (*P. Anderson-Witty*)

Adelie penguin colony on Signy Island (*P. Anderson-Witty*)

An enormous colony of emperor penguins at Halley Bay (*P. Anderson-Witty*)

Adults and chicks await the return of the other parent with food
(*P. Anderson-Witty*)

Emperor penguin parent with its chick resting on its feet
(*P. Anderson-Witty*)

Polar bears spend so much time in the water that they are considered to be marine animals (*Michael Bright*)

Pilot whales become stranded on Sable Island (*S. Anderson-Witty*)

Payne immediately recognized that certain phrases heard in the boat were the same as those on Watlington's tapes, and subsequently went on to reveal that the humpbacks were not making random sounds but regular repeating patterns that can be considered as true songs. They differ from bird song, though, in that a single song may last for as long as 30 minutes, and that song bouts may last all day. A whale recorded in the Caribbean by Howard and Lois Winn, of the University of Rhode Island, sang non-stop for 22 hours and was still going strong when the Winns pulled up hydrophones and sailed away.

It also turns out that all the whales in a particular area sing the same song, although those in the Atlantic sing a different song from those in the Pacific, and that the songs are continually changing. Whether a 'song-leader' determines when changes should take place is not known. A phrase may drop in pitch, a new segment may be added or sections might be shortened. New pieces are always sung faster and therefore at a higher pitch than older sections. The entire song is changed completely after about eight years.

Humpback whales sing mostly on their breeding grounds in the tropics and rarely on the summer feeding grounds. Only males sing, which gives us a clue to their function, if not the mechanism involved. It could be that the calls are given by male whales, both to space themselves from other calling males and to impress passing females. The area in which the whales sing might be considered as a gigantic lek. This is a territory held temporarily by a male for the purposes of displaying to the females. Several territories are often grouped together in the same area. Many animals making a lot of noise will attract others from considerable distances. Singing might be a means of announcing that a territory is occupied and so a way of spacing out the males. The dominant males, their place in the hierarchy perhaps determined by the sounds they make, sit in the central area where they are likely to be seen by the most females. As marine bioacoustics expert William Schevill once put it, 'The sonorous moans and screams associated with migrations of humpback whales past Bermuda and Hawaii may be audible manifestations of more fundamental urges . . .'

Peter Tyack, of Rockefeller University, New York, has spent many hours with singing whales and has studied their behaviour.

A singing whale is almost always alone, separated by several hundred metres from any other whales. It moves slowly, turning this way and that while singing. One underwater cameraman saw a singing humpback appear to dance with its song. Sometimes a singing whale is approached by a non-singer and the two will swim along for a short while, after which the whale previously singing swims off rapidly and silently as if displaced from its 'song-post'. The newcomer then sings in the other's place.

If a singing whale sees a cow and calf swimming by, it will turn and head directly towards them. The cow and calf will either continue to swim along in the same direction or move rapidly away, but never towards a singer. Those females that steer clear of singers may not be receptive. With a cycle where the cow may conceive one year, give birth the next, and raise and wean the calf the next, she may not be ready to mate in any one particular year.

If a singer catches up with a cow and calf, he stops singing and takes up station alongside the cow. The cow, calf, and escort move silently along, the calf keeping close to the female. Occasionally the cow and escort dive deep below leaving the youngster at the surface. Up to this point, things have been relatively peaceful, with the occasional flippering or rolling, but when the group encounters another singing male the trouble starts. The singer stops singing, races over, and swims along with the group as a secondary escort. Suddenly, their behaviour changes. They swim faster, and the secondary escort fights with the primary escort for the right to swim with the group. A whole barrage of sounds is now heard and the two males slap each other with their flukes, ram their bodies together, and blow bubbles. The commotion inevitably attracts more attention and still more singing whales stop their singing and join in the ruckus. The ever enlarging group becomes extremely rowdy and engages in a great deal of aggressive behaviour, each male trying to take the prime position next to the female. There may be as many as fifteen animals in such a group. After a while, the escorts appear to get bored and one by one they peel away until the cow and calf are left to swim on peacefully by themselves.

Lunge-Feeding and Blowing Bubbles

The most impressive of nature's spectacles, perhaps, is the sight of humpback whales feeding. On their breeding grounds in the tropics whales tend to avoid each other, but on their feeding grounds in sub-polar latitudes they cooperate in order to maximize their catch of small fish and krill. They have two main fishing techniques visible at the surface – lunge-feeding and bubble-netting.

In vertical lunge-feeding, the whales dive below a school of small fish and then shoot rapidly upwards with mouths agape. At the surface the agitated krill and small fish disturb the water, causing a hissing sound. Momentarily, all goes quiet; then, with an explosive roar, all the whales emerge at once, their throats distended with fish, krill and a huge quantity of water. They sink slowly back into the sea, pushing the tongue against the roof of the mouth to force the water out through the baleen and retain the fish and crustaceans, which are swallowed. Usually a maximum of seven whales lunge together, and they swim so closely together that they can be touching when they erupt in a 15m (50ft) diameter circle at the surface.

The lunge can be at any angle or simply across the sea surface. Horizontal lunges are sometimes performed with the whale upside down or on its side with a flipper, like a pennant, sticking out of the water. Several animals, swimming in line abreast, lunge-feed in synchrony – a process known as 'echelon' feeding, and in Glacier Bay, Alaska the same group of six individuals have been seen feeding in this cooperative manner for several seasons.

In bubble-netting, the submerged whales lie below the target school and concentrate the small fish, often herring, by surrounding them with a bubble curtain. One or two of the group rise slowly in the water blowing a narrowing spiral of bubbles. The herring flee from the flashing reflections and gradually form into a tightening mass in the centre of the column. The whales, with mouths open, push up through the centre of the bubble-net gathering in the fish. Just before the whales rise to the surface, a forty-five-second burst of trumpet-like sounds can be heard coming from one or two of the whales. Whether these calls are used to coordinate the efforts of the group or whether it is simply a dominant animal proclaiming its status, or even a means of disorientating the prey with sound, is not clear.

Recently, biologists watching whales off the Massachusetts coast have observed an individual using a technique different from any ever recorded, and from the fifty or so other whales feeding in the same area at the same time. It would rise out of the water with its mouth closed and then sink with it partly agape, apparently creating a sort of drain and catching the fish sucked down by the bathtub-type spout.

Deeper-Down Divers

The sperm whale is the deepest diving sea mammal. It has been established for certain, using underwater sonar, that individuals go down to at least 1,200m (3,937ft) and there is circumstantial evidence that they frequently go much deeper. A bull, thought to have been feeding on the bottom 3,200m (10,500ft) below the surface, was caught and found to have two bottom-dwelling sharks in its stomach. Several whales have been found entangled in submarine telephone cables hauled up from 1,140m (3,740ft), and many harpooned whales have old tin cans and other sea-floor debris, most likely shovelled up from the bottom, lodged in their stomachs. Why sperm whales should go so deep is not fully understood, although it is suggested that they are utilizing a food resource, mostly deep-living and bottom-dwelling squid, that is not exploited in a big way by any other creature. The seals, sealions, smaller toothed whales, dolphins, and a host of the larger fishes, skim off the bulk of the squid in the top layer of ocean, while the sperm whales dive below and avoid the competition.

As far as researchers can ascertain, sperm whales dive almost straight down at speeds up to 170m (558ft) per minute. Bulls dive the deepest and the longest, sometimes being down for up to two hours. Cows level out at about 1,000m (3,281ft) and stay down for an hour, while youngsters reach about 700m (2,297ft) and come up after a half-hour at the most. They return to the surface almost as fast, and have the same mechanism as seals for avoiding the 'bends' (see page 27), except that the absorption of oxygen by the muscles is even more efficient than that in other sea mammals.

A whale might go down many times with only a few minutes between dives, although it must eventually stay and rest at the surface. It is able to descend and ascend rapidly using its powerful tail for propulsion, but aided by a unique buoyancy mechanism in the head.

The sperm whale's characteristic shape is the result of a large mass of tissue, known as the spermaceti organ, at the front of the boiler-shaped head. It is filled with a special wax – used by man in former days for high-quality candles and in cosmetics – that has the property of melting at 29°C (84.2°F) precisely, and is permeated with a complex plumbing system of blood vessels, sinuses and nasal passages. By causing the wax to melt or solidify in a controlled manner, according to Malcolm Clarke of the Marine Biological Association of the United Kingdom in Plymouth, the whale is able to control its density in the water and affect its ability to go down or come up.

At the surface, seawater is circulated to cool the wax, which shrinks and becomes more dense, causing the whale to sink. At the end of the dive, body heat generated by the muscles is carried by the blood into the spermaceti organ where it melts the wax, causing the head to be less dense than the water, and allowing the animal to rise to the surface with the minimum of effort. No matter how exhausted the animal becomes at the bottom it is assured a safe journey to the surface – important, perhaps, after a long feeding expedition or a battle with a giant squid.

Killing with Sound

Dolphins bounce high-frequency sounds off objects in their environment in order to locate their position and determine their nature. The sounds are produced in the complex plumbing system of nasal sacs and plugs below the blow-hole at the back of the head and are focused in a 'fatty' organ known as the 'melon' in the bulbous forehead. Much as light is focused through a lens, the dolphin concentrates ultrasound into a narrow beam which is projected in front of the animal. The sound beam can be of such an intensity that, if it were louder, it would turn to heat. If such a powerful beam were directed at a prey animal, contends Ken Norris, of the University of California at Santa Cruz, and Bertel Mohl, of Aarhus University, Denmark, then it could be knocked out or even killed.

This idea is not new. Soviet workers have calculated that the dolphin's echo-location beam should have enough energy to debilitate prey. A.A. Berzin found evidence that the sperm whale, the largest of the toothed whales, might use sound to knock out giant squid. Working on whale factory ships, he found whales hauled out with congenital abnormalities of their jaws, yet with stomachs full of squid. How did they catch their prey? Squid, after all, are fast jet-propelled animals that surely could outrun bus-shaped sperm whales. Recently, British and American researchers following sperm whales in small boats have noticed very loud 'pistol' shots amongst the general clicking cacophony emitted by a foraging group. It could be that they are hearing the powerful knock-out sounds.

In the laboratory it was shown that fish could, indeed, be killed with high intensity sound beams, and some experiments with dolphins in captivity revealed that the creatures not only have the capacity to stun with sound but that they do so. Live fish were placed in a large tank with three Hawaiian spinner dolphins. The dolphins sprayed the shoal with sound and gradually the fish became disorientated. In the wild, striped dolphins have been seen to echo-locate a school of anchovies, and then cut through the middle of them with mouths wide open, shoveling the fish in at will. The anchovies would make no attempt to escape.

If the interpretation of the events is correct, then dolphins clearly have a formidable weapon in their head. In a large school of dolphins, though, how does an individual ensure that it does not zap another? It seems that dolphins have 'echo-location manners'. Listening to and watching the three dolphins in the tank, Ken Norris found that not once did a dolphin direct its sound beam at one of its companions. Every time another dolphin passed in front of an animal which was actively locating, the latter would switch off, turn its head, and then resume echo-locating elsewhere.

Orca: a Whale Called Killer

For centuries man has been in awe of the orca. The Romans named it after the king of the Netherworld; the Germans call it the 'sword whale' – a reference to its tall sword-shaped dorsal fin that can stand 2m (7ft) above the surface of the sea; the Scandinavians know it as the 'chopper of blubber'; and in Spanish it is the 'assassin'. In English it is simply the killer whale – a named derived from a savage reputation as the only whale to devour warm-blooded prey – behaviour recognized and feared by early whalers, and recorded in an essay to the Royal Society in 1725 by the Honourable Paul Dudley.

> They go in company by Dozens, and set upon a young Whale, and will bait him like so many bull-dogs; some will lay hold of his Tail to keep him from threshing, while others lay hold of his Head and bite and thresh him, till the poor Creature, being thus heated, lolls out his Tongue, and then some of the Killers catch hold of his Lips, and if possible of his Tongue; and after they have killed him, they chiefly feed upon the Tongue and Head, but when he begins to putrify, they leave him.

Although give the name 'whale', in reality the orca is not one of the great whales, but the largest, fastest, and most powerful member of the dolphin family – a supercetacean with peg-like teeth. An adult male can grow to 9m (30ft) long and weigh 9 tonnes – the length and weight of a single-decker bus. It is one of the swiftest animals in the sea, able to overtake and outmanoeuvre its prey – the sea's supreme predator.

'Like man', wrote one modern marine biologist, 'orca rules its domain, or did, until recently challenged by man. Man and the killer whale are the two most formidable, successful and intelligent social predators ever to live on earth.'

Orcas are found in every ocean and sea in the world – polar, tropical and temperate – but perhaps the greatest concentration of orca *watchers* is along the Pacific coast of North America, particularly the waters around Vancouver Island. At the Pacific Biological Station at Nanaimo Mike Bigg has been studying orca social behaviour. Using a photographic technique to identify individual whales by their markings, he has found that they live in stable social groups, known as pods.

A pod is a group of ten to twenty individuals, probably related to each other through the mothers, that often travel and hunt together. There are usually two or three mothers and a granny or two that dominate the group, and a number of juveniles, youngsters, and bulls. The bulls, though much larger sometimes than the cows, do not seem to have the same social status as the older mothers. A young male, as he grows, stays with his mother. It is very much a matriarchal society. Mike Bigg recalls seeing a bull lifted half out of the water by an angry cow, even though the bull was much larger.

A pod, however, does not always travel as one, for sometimes small sub-groups will split off and travel separately from the main group, perhaps to hunt and feed. The role of these sub-groups is not clear, but eventually they return to the fold, and so, in the long term the main pod remains stable.

Early in the study, Mike Bigg discovered that two kinds of killer whale pod live in the waters around Vancouver Island. There are those seen for the best part of the year, and labelled 'resident' pods, and those spotted infrequently, which became known as 'transient' pods – it was as if they were in transit from one place to another. Further research revealed that 'residents' and 'transients' were quite different, both in appearance and behaviour – two forms genetically isolated and not interbreeding.

The 'resident' kind of killer whales, characterized by a slender, rounded dorsal fin, travel in large groups, sometimes as many as fifty whales in the group, and appear around Vancouver during the summer months when the Pacific salmon is returning to its home river to spawn. 'Residents' follow the salmon migration routes, so their movements are predictable. Their lifestyle seems to be determined by the behaviour of their prey. Salmon swim in large, mobile shoals. It is a food source that is said to be 'clumped'. In order to detect and locate a 'clump' it is advantageous to have a large group that can spread out and scan a wide area.

'Transient' whales, characterized by a stubby, triangular dorsal fin, are less predictable and travel closer inshore to feed on more thinly dispersed and less abundant food sources. They take marine mammals, such as seals and sealions. In the Vancouver area there are only 50 'transient' whales, whereas there are 250 'residents'. There is some speculation that the two forms might represent different species. Soviet workers have described similar characteristics in two distinct killer whale types in the Antarctic. That investigation is still going on, and is, as yet, unresolved.

Around Vancouver Island, it has been found that the 'resident' pods live in two large communities – the northern community around Johnson Strait and the southern community. Between them lies a no-go area. The two communities never mix. To recognize the two distinct populations, not only do the researchers rely on the individual markings, but also on the remarkable sounds that the killer whale makes. John Ford, also at Nanaimo, has been listening in and trying to interpret their meanings.

Killer whales make clicking sounds that are used for echo-location and a variety of social sounds, such as squawks, burps and squeaks. The social sounds are of two forms: those that are variable and complex and which are used when the animals have congregated and are 'chatting' together in subtle ways that we are far from interpreting; and also those sounds that are discrete, stereotyped, and repetitive, calls used by all the animals in the group, when they are on the move. They are contact calls, used to keep the animals in touch when the pod is scattered. Often an animal will emit a call and all the others in the group will respond and repeat it. As yet, John Ford is unable to link calls to any particular pattern of behaviour – there is no specific meaning.

In general, there is some relationship between the frequency with which calls are given and the level of activity of the pod. When the whales are excited, the numbers of calls increases, although when a 'resident' sub-group has cornered salmon they tend to stop calling. When they have caught a fish they give a few calls and then catch up with the rest of the group. 'Transient' groups, on the other hand, tend to forage in total silence. Seals are also creatures alert in the auditory domain and would easily detect the presence of a noisy whale. 'Transients' are therefore sneaky in their approach.

Most of Ford's communication research has been with the 'resident' pods of Vancouver Island's northern and southern communities, and he has found that each pod has its own readily identifiable calls – pod dialects. Even the untrained ear can hear the differences between some of the pods, but perhaps the most exciting find is that some resident pods have call types with striking affinities to the call types of certain other pods. In the northern Johnson Strait community there are 13 pods that can be divided into three 'clans' that share sounds. It is likely, thinks Ford, that pods sharing calls have descended from a common ancestral group and as it grew and split, each of the new pods spent more and more time in social isolation and the dialects of the new groups gradually drifted apart. Pods within a 'clan' share a number of stereotyped calls, but pods in different 'clans' have no calls in common. So, instead of just one ancestral pod giving rise to the northern community, by splitting, it might have been three ancestral pods that entered the area at different times. The largest 'clan', with eight pods, may have arrived earlier than the two smaller 'clans' with only two and three pods.

Dialects are rare in mammals for vocal learning is unusual – the sounds made by most mammals seem to be innate. Among the mammals, the only groups that are known to have a vocal learning system are the whales and dolphins, and man. And of those, it is only man and killer whales that have dialects among neighbouring groups that also interact.

The most frequently documented aspect of killer whale life is that of hunting and feeding, and again the 'resident' and 'transient' pods of Vancouver show different strategies. Although members of 'resident' pods travel and locate fish schools across a broad front, they do not herd fish or hunt cooperatively. When a fish is detected ahead, an individual will go for it alone. A 'transient' hunt is far more dramatic. A small pod will take two to three hours to butt and beat their tails against a large seal or sealion until it is unconscious. Then they take it below the surface, drown it, and then consume it. The 'transient' killer whale, despite its size and power, is a cautious predator that does not take any chances that it might be injured by a bite from its prey.

In the South Atlantic, killer whales have been seen to swim right up on to the beach, pluck off a fur seal, and return to the water on the next wave. In the Antarctic, they swim up to an ice-floe, on which penguins or seals are resting, and bump it so that the animals fall into the water. Sometimes they shoot straight up through thin ice to grab their prey.

In different parts of the world, food preferences and availability can vary considerably. Off Japan, cod is supplemented by halibut; near the Brazilian coast the diet is brightened by the occasional sting ray; and in the northern Pacific, sealions and elephant seals are taken alongside dolphins and porpoises. Frank Robson remembered an organized attack on a school of dolphins off the coast of New Zealand.

> The killers must have heard their chatter and closed in silently, intent on the kill. They had formed a circle round the dolphins and then they moved steadily in, closing the circle, driving the hapless dolphins before them into a crowded ring. And, as they closed, they kept up their terrible bleeping . . . When the Killer circle had decreased to about fifty metres across, with the dolphins swimming round nose to tail, three or four of the hunters entered the enclosure and selected as many victims as they needed to satisfy their hunger. These they maimed, biting some across the tail so they could not swim away.

The most dramatic attacks observed have been those involving the great baleen whales. Hunting groups of orcas follow the migrating grey whales along the Pacific coast of North America. Their targets are the young and the infirm. Around Newfoundland and off the coast of Alaska humpback whales, feeding on krill and small fishes, are pursued mercilessly until weak animals succumb. In 1977, a whale researcher watched a pack of killer whales that was engaged in a five-hour-long attack on one of the largest creatures on earth – the blue whale.

> The predators exhibited marked divisions of labour. Some flanked the blue on either side, as if herding it. Two others went ahead and two

stayed behind to foil any escape attempts. One group seemed intent on keeping the blue underwater to hinder its breathing. Another phalanx swam beneath its belly to make sure it didn't dive out of reach. The big whale's dorsal fin had been chewed off and its tail flukes shredded, impairing its movement. The bulls led forays to pull off huge chunks of flesh.

Miraculously this whale survived the onslaught, but it is likely that it died later from its wounds.

Perhaps the most bizarre killer whale story is of an orca that was lost some 80km (50 miles) up a Scottish river: it took to catching ducks. 'The whale would spot a duck and start after it, the duck taking flight when it saw the large dorsal fin approaching. The ducks were unable to gain altitude quickly enough, and were snapped up while they were flying with their wings pattering on the water as they tried to escape.'

Despite its reputation as an indiscriminate killer, there are few records of killer whales attacking man. On Scott's polar expedition to the Antarctic in 1911, the photographer H.L. Ponting was nearly knocked from an ice-floe, probably mistaken for a seal. The only story of an attack that has passed into the literature is from British Columbia.

Two lumberjacks were rolling logs into the water to build a log raft, when a pod of killer whales swam by. One of the men deliberately hit one of the whales with a log. The pod moved away and disappeared. Later that evening, the lumberjacks were rowing back to their camp across the bay when the pod reappeared. They bumped and overturned the boat, and the man who lobbed the log was never seen again. His companion swam ashore, unharmed, and so lived to tell the tale.

For killer whales, though, the danger is often reversed. Macho sharp-shooters take pot-shots at orcas swimming close to the shore, and in some parts of the world orcas are taken alive for a thriving multi-million dollar trade in captured animals for dolphinaria. But, for every whale in captivity – usually a youngster for reasons of economy and space – it is thought many must have died, some from less obvious causes such as frostbite while waiting in holding pens. With the birth rate of captive orcas close to zero, and a very short life expectancy for performing adults, it is clear that the only way to restock dolphinaria is from the wild. The morality of doing so is being increasingly called into question. Should seemingly intelligent marine mammals be kept in captivity at all? Are we condemning them to a life of misery and boredom and, in the case of performing animals, are we demeaning them? The scientific community is divided. Some claim that it is important to have animals in captivity for research and education. Others believe that these creatures should be in their rightful place – in the sea, in the wild.

Whale Strandings

In September 1983 a herd of 80 pilot whales found themselves stranded on Tokero Beach, Northland, New Zealand. Local people, some of whom had read about humane ways to save stranded whales, waded out into the shallow surf to help them. They talked quietly to the whales and kept their skin wet until the tide turned and there was sufficient water to refloat and turn them. One by one the whales were helped out into deeper water. Then a surprising thing happened. A school of dolphins, which had been feeding nearby, swam in amongst the whales and led them out to sea.

This was not the first time that dolphins have been observed to assist whales. In 1979, a large herd of pilot whales went aground in Whangarei Harbour, to the north of Auckland, and after people had got them swimming again a dolphin school guided them to the safety of the open sea. How, though, did the dolphins know that the whales were in trouble? Is there some form of interspecies communication which we know nothing about?

One mystery that may have an answer now, is why the whales became stranded at all. Whales often accidently beach in roughly the same places – 'accident-black-spots' for whales. These stranding sites are thought to be geophysically interesting, for they are places where lines of geomagnetic variations, known as 'magnetic valleys', cross the coastline or are blocked by islands. The suggestion, from Margaret Klinowska, of Cambridge University, is that whales are using geo-magnetic cues, much like an automatic pilot, to guide them through the sea. They cruise along with their other sensory systems, like vision and echo-location, switched off, and if they reach a magnetic valley, which happens to coincide with a sand bank or a beach, they are caught unawares. Suddenly they find they are in shallow water, and, dis-orientated and confused, they continually swim on to the shore.

There is some circumstantial evidence to support the idea. There are more strandings of deep-water species than of inshore species. Most coastal water dolphins die of natural causes and are washed up on beaches. Only 6 per cent of live strandings in British waters, for example, have involved harbour porpoises. The figure is 67 per cent for ocean-going false killer whales. Could it be, then, that the New Zealand dolphins were using their local knowledge to guide the pilot whales to open water where they could continue their journey on a magnetic map?

A magnetic sense for cetaceans was suggested when Californian scientists discovered particles of magnetic material, associated with a net of fibres, in the brains of several Pacific dolphins and a beaked whale that had been washed ashore on Californian beaches. The association of the fibres and the magnetic material suggested to the researchers that the magnetite was not a bi-product of metabolism but must have a function in dolphin life. They speculated that the tissue is, indeed, a magnetic field receptor and could be aiding dolphins and their cousins, the whales, in orientation and navigation.

4
Europe

The outline of Europe is the result of a sinking continent. The tongue of Norway and Sweden, and the bulge of France and Spain is all that is left of a once larger land mass that ran along the western side of the great Eurasian plain. The north is the product of ice action – U-shaped valleys, ribbon lakes, fjords and smoothly sculpted rocks, whilst the south is the result of mountain building, as Africa pushes north and twists clockwise against the geologically unstable Mediterranean area. In between, is the plain of central Europe.

From the great northern forests of Scandinavia to the hot and dry olive groves of the northern shores of the Mediterranean, there are a multitude of different habitats in which a great diversity of plants and animals can live – moorlands, freshwater and saltwater marshes, estuaries, deciduous and coniferous forests, meadowlands, mountains, rivers and lakes, streams and ponds, grasslands, dunes, gardens and cities.

The Large Blue and the Little Red

The large blue butterfly is thought to be extinct in Britain but its demise has been the result of an unfortunate series of environmental events and an extraordinary relationship with red ants.

The life history of this butterfly was unknown until 1915 when Bagwell Purefroy and F.W. Frohawk tried, but failed, to breed it in captivity. The adult female laid her eggs on thyme buds and, a week later, the tiny caterpillars hatched out and fed sometimes on each other but more often on the thyme plant itself. When they changed their coat for the third time, becoming what is known as third instars, they simply stopped feeding and died. Purefroy began to unravel the mystery when he noticed a red ant carrying a caterpillar back to its nest. Seventy years later, entomologists such as Jeremy Thomas, of the Institute of Terrestrial Ecology at Furzebrook in Dorset, are still unravelling the remarkable story.

In the autumn, when the large blue caterpillar reaches its third instar, it changes from a herbivorous diet to a carnivorous one. It stops eating the thyme leaves, climbs down from the plant, and wanders around on the ground until it bumps into a red ant of the species *Myrmica sabuleti*. Instead of stinging or biting the caterpillar the ant becomes very excited for the larva rears up and produces a drop of 'honey' from a special gland on its abdomen. The sweet liquid is attractive to the ant and it grasps the caterpillar in its jaws, and carries it right into the heart of the ants' nest. The other ants in the colony do not attack it, as they would any other intruder, but are appeased by the production of a sugary solution on the caterpillar's skin. The ants crowd around and 'milk' the caterpillar, palpating it with their antennae to stimulate more secretions. Pheromones are also released by the caterpillar to suppress the natural tendency of the ants to tear the caterpillar to pieces. In this way the caterpillar is afforded some protection in the ants' nest and can freely go about its business of fattening-up for the next stage of the life cycle. It repays its hosts by gobbling up their larvae. The ants, however, still do not harm it.

The caterpillar continues to live in the nest, eating the ant larvae and pupae, for about five to six weeks. It then hibernates in the ants' nest until spring, when it changes into a pupa or chrysalis. As a pupa it continues to placate the ants by producing sugary solutions and pheromones, and in addition it makes peculiar rasping sounds by scraping the tip of the abdomen against the pupal case. It turns out that these noises are the same as the sound communication signals being exchanged between the ants themselves. Ants are able to 'talk' to each other in a variety of ways including by smell, taste, vibrations, and sound. The sounds are produced when the back legs are rubbed together. So the large blue chrysalis copies not only the smell and taste signals, but also the ants' complicated sound system in order to live safely in the nest. And the story does not end there.

Towards the end of the pupal stage the chrysalis becomes particularly noisy, causing many ants to gather around. When the adult butterfly emerges the ants do not attack it for it too produces secretions that both excite and pacify the hosts. The adult then heads for the surface escorted by the ants. Outside it would be vulnerable to small predators, such as ground beetles, for it must rest on a plant stem until its wings are expanded, but by continuing to produce secretions it retains its escort which will ward off any attackers.

With such a complicated and precise life cycle the large blue butterfly is helpless when drastic changes take place in the environment. Large blues are usually found on chalk downland where the grass is kept short by the activities of rabbits and where thyme plants are abundant. These are the conditions essential for the survival of the red ants. When the rabbits succumbed to myxomatosis and all but disappeared from the land, the grass grew longer, the ants disappeared, and so did the large blue butterflies. Recently there have been several attempts to breed large blues in Britain using stock imported from the continent where the butterfly is still common. Unfortunately, so far the experiments have failed.

Cheating the Cushion of Death

Sundew plants of the genus *Drosera* grow in boggy or sandy places where nitrogen is in short supply. To supplement their nutritional needs they have turned to trapping and digesting insects. They are carnivorous plants. The leaves are modified to form round or elongated pin cushions from which protrude many stalked, glandular hairs. Insects, attracted to the leaves, are caught on the hairs and trapped by the sticky secretion. The struggling movements of the prey cause the outer hairs to bend into the centre, thus preventing the insect from escaping. Enzymes are released that digest the tissues of the victim. When the process is completed the hairs and the leaf open back once more, ready for another visitor.

The caterpillar of the plume moth *Trichptilus parvulus*, however, is able to by-pass the sticky hairs and feed on the plant without coming to any harm. The caterpillar becomes active at night when it eats the glandular hairs. It starts from the middle and works its way to the outside, so gaining a measure of protection from predators from the larger hairs around the periphery of the leaf. It has the minimum of contact with the hairs by only touching them with the long bristles that cover its body. When it is moving the bristles are easily withdrawn from the sticky secretions.

The adult moth coming to lay its eggs on the sundew also has a trick or two up its sleeve, or rather on its wings. Plume moths have feathery edges to their wings. These are lined with detachable scales that can be shed if the moth accidently brushes against the hairs. They are very useful too if the moth should find itself blundering into a spider's web.

Spider Bites Man

In the British Isles there is little danger of being bitten by a black widow, a brown recluse or a funnel-web, all spiders with venomous and potentially fatal bites, for they live elsewhere in the world. Indeed, it was thought that there was no danger of being bitten by a British spider at all – that is, until Eric Duffey and Roger Plant, of the Institute of

Terrestrial Ecology at Monks Wood, conducted an informal national survey in 1981, identifying the spiders from their victims' descriptions.

They discovered that even minute 'money' spiders were capable of biting into human skin and were a particular nuisance at sewage works. The majority of bites from larger spiders, though, occurred in the house, followed by gardens, garages and the countryside. Most bites were on fingers or legs, but just about any part of the body is vulnerable, including the buttocks.

One lady was bitten on her bottom by a spider when she got into bed. Her husband was so incensed that he despatched another potential offender with a blast from his shotgun. The most likely hiding places from which a spider would emerge to bite turned out to be the folds of clothes, trouser legs, and bed clothes. The number of species responsible read like a *Who's Who* of British spiders – the garden spider *Araneus diadematus* and the black-bodied and white striped *Araneus umbraticus* were found to bite the occasional gardener; house spiders *Tegeneria sp.* were thought to be the main culprits indoors, along with the chocolate-brown *Steatoda bipunctata*, the sinister-looking mesh-web spiders *Amaurobius sp.*, the red-legged *Dysdera sp.* and several others that frequent country homes especially. Bites ranged from pin-pricks without any skin reaction to quite painful bites that swelled-up and stayed swollen for several days. A bite from one of the largest of the British spiders, *Clubiona corticalis*, was described by one woman as being like a wasp's sting.

The Wild Cattle of Chillingham

Over seven hundred years ago a herd of cattle grazed the Cheviot Hills. Today, the descendants of that herd are to be found, roaming almost wild, in Northumberland. They represent one of the last herds of horned white cattle surviving in the British Isles. Others are to be found at Cadzow Park in Lanarkshire and Chartley Park in Staffordshire – attempts to disperse the herd and eliminate the risk of having to destroy the entire stock if there was an outbreak of foot-and-mouth disease.

The Chillingham cattle are primitive domesticated animals, unaltered by cross-breeding or imported animals since medieval times. They are also semi-wild and so give us some insights about the way their truly wild ancestors behaved.

Smell is the predominant sense. Hearing and sight (domestic cattle seem unable to detect the red end of the colour spectrum) are adequate but not as refined as in hunting animals. Cattle are able to smell an approaching predator, identify the best vegetation to eat by its aroma, and communicate across the herd with scents containing chemical messengers (pheromones).

Chillingham bulls, unlike their fully domesticated relatives, are able to pick out a receptive cow and isolate her from the rest of the herd. They will undertake ritualized fighting, much like that of the North American bison, for dominance in the herd and for the receptive females. A female is briefly bonded to a bull after a fighting bout, and if he is not displaced during another bout, he mounts her.

Chillingham cows, like domesticated cows but unlike other wild bovids, such as bison, move away from the herd to give birth, which they do standing up. The afterbirth is eaten and the calf licked all over to minimize the risk of disease and infestation by flies. The licking also appears to stimulate the calf to defecate. Very soon after the calf has dropped to the ground it attempts to stand and feed from the udder. Curiously, during the first four days of its life, the mother leaves the calf after feeding and returns to the herd. At regular intervals she will go back to the calf, which lies hidden in the grass, feed it, and return once again to the main herd. On the fifth day the calf follows its mother everywhere, and at first becomes the centre of attention, much to the consternation of the mother which will chase away inquisitive cows.

The Chillingham cattle are aggressive animals and do not take kindly to intruders. If approached, the herd forms into a circle with their horns facing outwards. Move closer, and they will charge. The bulls are particularly bad-tempered.

Today they are left to their own devices. Nobody interferes with them. It is said that if a calf is touched by a human helper it will be killed. They look primitive, with their square heads, stout curled horns, and powerful bodies, but they differ considerably from the wild ancestral ox – the aurochs – from which they, and all domestic cattle, have descended.

The aurochs was, at one stage in its history, an enormous and formidable beast. The atlas bone, at the top of the vertebral column, from an aurochs skeleton, found in caves at Charterhouse under the Mendip Hills, indicates a beast twice the size of a Chillingham bull. But it was not always that large. The aurochs were medium-sized herbivores at the onset, gradually evolved to giants, and then declined once more to cattle-size. In their heyday they were distributed across the entire Eurasian continent from the Atlantic to the Pacific. From the cave paintings of Lascaux in southern France we known that the females were a reddish colour and the bulls were larger and darker. They arrived in Britain during the Hoxnian warm interglacial period and were either hunted or domesticated.

Before domestication, the aurochs was thought to have been a religious symbol of strength and virility. Aurochs horns have been found in Turkish shrines dating from 6500BC. The first recognized site of domestication is at Nea Nikomedeia, a neolithic village in northern Greece.

As recently as the eleventh century they were still common throughout Europe, but in the thirteenth century there were signs that they were on the way out. The last individual died in a paddock in the Jaktorow Forest in Poland in 1627.

Attempts have been made to breed back, by interbreeding various primitive cattle, the characteristics of the wild aurochs. In West Germany a zoo has recreated animals that closely resemble the animals depicted in seventeenth-century paintings.

The Wild Dogs of Italy

Since the Second World War, hunters and farmers with increasingly sophisticated firearms have reduced Italy's wolf population to about a hundred individuals. Most of the remaining animals live in small groups not far from Rome in the Abruzzo National Park, but despite their legal protection their future survival is threatened not directly by man, but by an animal that has taken over their ecological niche – the feral dog.

Feral dogs are domestic dogs that have escaped or have been left to fend for themselves. They live in the wild, scavenging on scraps of food found at waste dumps or in garbage cans. Some have grouped together into packs and are reverting to behaviour normally associated with their ancient ancestor, the wolf, and by interbreeding with Italy's few surviving wolves may eventually bring about their extinction as a pure species.

One of the major problems is that there are so many. A count a few years ago came up with an estimated 80,000 feral dogs, and that does not include urban strays. And they could be dangerous, for many of the wild dogs are large – German shepherds and Pyrenean mountain dogs. Sheep are definitely at risk, but so are larger animals. Geoff Carr, of Oxford University, who has been studying the behaviour of feral dogs, found that they will readily take an animal as large and as powerful as a horse: 'There was one group of dogs in the area where I was working which seemed to specialise in taking horses. Although they could not catch horses that were free to run, those fenced in a field and couldn't escape were brought down.'

The dogs taking farm livestock tend to live in the remote and sparsely populated parts of the Appennines – the hill and mountain country normally inhabited by the wolves. Natural prey, such as deer, has already been exterminated by human hunters. So, both dogs and wolves come down to where people are living and turn to garbage for their daily food. Many of the villages have garbage tips, and there are often piles of debris littering the sides of the roads. Now, both animals rely on a 'clumped' food source. This means that many animals can feed at the same sites. There is no competition as there is plenty of food at the tips.

The Italian feral dogs can be divided roughly into those that live mainly in the mountains and run around in packs like wolves, and those that frequent the valleys, probably recent arrivals, and forage alone like foxes. Several loners may amicably share a home range area but will aggressively defend it against any newcomers or animals in a neighbouring patch.

'We can see that these animals are defending territories', concludes Geoff Carr. 'The dogs that live in neighbouring territorial areas know the boundaries and do not often overstep the mark. They know that they would have to leave if they are discovered as intruders. They would be chased out. Sometimes, though, foreign dogs, that do not know the rules, wander deep into a territory and if they get caught they most likely will be killed.'

One of the major problems with wild dogs is their relationship with wolves. Under normal conditions, if a pack of wolves met a pack of dogs, the wolves would attack and kill the dogs. But there are so few wolves that a young male leaving his pack to find a female and start his own family is more likely to take up with a feral dog than a wolf. Thus, his 'wolfness' will be diluted, by the contribution of genes from the feral dog, in the next generation, and the genetic purity of the wolf population is compromised. Luigi Boitani, of Rome University, remembers seeing the litter from a mixed mating: 'One she-wolf that mated with a shepherd dog had six pups – two were wolflike but the other four were black with a white front leg.'

Another problem is that the dogs are breeding more frequently than the wolves. Dogs, which are thought to have descended from wolves, have been bred to reproduce more often. A wolf pack will only contain one breeding female, and she will mate only with the dominant male. In dog packs, several females will reproduce and they mate with more than one male.

With so many dogs and so few wolves, livestock killings are more likely to be the result of attacks by feral dog packs, but it is the wolf that still has the evil reputation and it is the wolf that still gets the blame.

'Dogs have a huge evolutionary advantage', says Luigi Boitani. 'Who would scream at a dog when it comes close to livestock? Nobody. If it were a wolf, everybody would scream and probably shoot. A dog would have to kill before it caused alarm, and that is certainly an advantage for the dog.'

More confusion arises in the way that wolves and dogs kill their prey. When in a field full of sheep both will indulge in 'surplus killing', that is they will seemingly go berserk, killing as many animals as they can – far more than is necessary to satisfy their immediate needs. David MacDonald, of Oxford University, points out:

Many carnivores will do this, but it is more notorious for creatures such as foxes, dogs and wolves. They get into circumstances where prey, for reasons better known to the prey and obscure to the carnivore, refuse to run away and the carnivore just continues to kill. In the case of Italian livestock, they do not run away because they are fenced in. Of course, evolution cannot be expected to take into account the fencing of sheep when it was designing predators. If the prey's escape behaviour is compromised, the predator behaves as is normal – it takes every opportunity it can because it will not get too many opportunities in the natural world – and grabs as many sheep as it is able. The difficulty then is that the shepherd loses too many sheep.

And this presents a further problem for the wolf. Interbreeding with dogs has meant that it is not clear to the local villagers whether the culprits are dogs or wolves, and so any action taken is likely to involve both. Even if the shepherd is conservation conscious, he is going to find it difficult to distinguish dog from wolf or wolf-dog crosses.

'The feral dog problem in Italy is a conservation nightmare', suggests David MacDonald, 'for any action taken against dogs is bound to affect wolves, and wolves are much more susceptible because they are so thin on the ground.'

The Serpents' Year

The adder is not immortal. It does not swallow its young. And, it certainly does not spring spontaneously from a horse-hair dropped on to cow dung. It is, however, Britain's only venomous snake, and one with a remarkable life history.

At around the end of February or the beginning of March, but always a fortnight before the females, the male adders emerge from hibernation. The ground temperature reaches about 8°C (46°F), and the torpid males, maybe twenty or thirty in one site, follow the gradient of warmth from their hibernaculum – the vipers' den – to the surface. There they lie about, sometimes in great balls ('plagues' of over a hundred snakes have been known), basking in the sun and gathering energy to complete the development of the sperm, and to make the other chemical changes necessary before they can slough their skin.

The first sign of this is a clouding of the eye, followed by a tear in the old skin along the corners of the mouth. It gradually peels off in one piece, catching on stones and twigs as the snakes move along the ground. The ancients thought that the snake was immortal, that at each slough it left behind the vestige of its previous life, and never died.

When females appear, courtship starts immediately, with males seeking out the females in response to a pheromone. Until now, reptile skin was thought to be continuous and watertight, unlike the porous amphibian skin, but scanning electron microscopes have revealed that there are soft, thin folds of epidermis between the hard, keratonized scales. Bumps in the soft strips turn out to be large, spongy-looking cells with numerous pits, and sections reveal them to be the openings of microtubules. It is thought that these are the release points of the sexually attractive chemicals. Swabs taken from courting females and smeared on snake models attract males from some distance.

The combat that these chemicals inspire in the males leaves the victor dominant over the vanquished for about 24 hours, during which time any chance meeting between the two will result in the weaker one being chased away. On some occasions, when the adders are the same size, the fight can be quite prolonged, and a smaller individual can sneak behind their backs and mate with the contested female. A dense population of captive male adders deprived of a female will not fight, even in the breeding season. But it is not smell alone. Males exposed to the pheromone without the appearance of a female still will not fight.

Combat complete, the winning male moves over the back of the female, nodding his head and flicking his tongue in and out to taste or smell her, and constantly tapping her with his chin. This characteristic jerky action can go on for hours or sometimes days, depending on the temperature. When ready to mate, the female raises her tail, exposing the cloacal opening, and waggles it back and forth. The male, meanwhile, gradually wraps his tail around the female's body in order to gain entry to the cloaca with one of his hemipenises. Male adders have two, one each for left-hand and right-hand coiling. The hemipenis has small soft spines, making uncoupling difficult, and when the sperm transfer is complete, the female moves off, dragging the still-attached male along with her. He goes willingly, because if he extended his scales to catch in the undergrowth the organ would be snapped off. Eventually, and much to the relief of the male, they disengage.

With courtship and mating complete, the snakes enter a feeding stage. Adders do not see well over distance but respond to movements, and apparently do not hear at all. Their main prey-detecting sense appears to be smell, and the flickering forked tongue samples the air for the tell-tale molecules of food. When an adder locates a small mammal, it strikes with its mouth agape, plunging its fangs into the animal's body and injecting a lethal dose of venom. Then, the snake 'yawns' – a manoeuvre to fold the fangs back into their normal resting place. The poison works fast, and an adder can wait until a mouse is fully inactive before eating it, since a mouse bite can kill a snake. As enzymes in the venom begin to digest the mouse's tissues, the prey is engulfed head-first and whole. The snake's jaws dislocate, and its skin and body stretch as the corpse is worked down past the backward-facing teeth and into the stomach. A single mouse is enough to keep an adult adder going for a couple of weeks, but it has been known, during an abnormally cold summer, for a snake to last for an entire year on just one meal.

In a dense population of adders, competition for food can be fierce. Again, combat is used to settle disputes. A large struggling mouse, in its death throes from the action of the venom, will attract not only the snake that attacked it, but any others in the vicinity. Ownership is established by a dance almost identical to that between sexually aroused males, but now it can involve females and juveniles as well. If a small snake is challenged by a larger one and does not let go of the prey, it can easily be gobbled up too. The females need nutrients for their developing embryos and, being larger and heavier than the males, are more likely to win bouts of 'predatory combat' and obtain the food. When times are hard this feeding strategy favours females and ensures that the offspring will survive at least to hatching. There is also evidence that when food resources are low, the snakes will migrate away from the hibernation area in search of fresh supplies.

After mating towards the end of May, the female sloughs and becomes a mobile incubator, basking in the sun, often on the same site each year, and warming the eggs until they are ready to hatch. Some climb trees in order to gain the last watery rays of the evening sun. Up to a dozen eggs are retained inside the female's body, each surrounded by a transparent membrane. No protective shell is needed. At this time female adders frequently fall prey to man.

On their basking sites they are more noticeable and are often needlessly killed. This has given rise to an old wives' tale which, strangely, gained a degree of scientific respectability. Female adders, so it says in the literature going back to Elien the Sophist in the second century AD, swallow their young when threatened. What modern herpetologists believe has happened is that when a spade has been brought down and viciously smashed a female's body, some of the youngsters have burst out from inside her and wriggled away. Eye witness reports from no fewer than seventeen respected English naturalists, of baby snakes entering the mouth of the female, have further enhanced the tale. It seems likely that they were witnessing babies hiding *underneath* the mother, a behaviour often seen in captivity.

Spontaneous generation once accounted for the birth of baby adders. Falling horse-hairs, so it was said, turned into tiny snakes on landing in a cowpat. In reality, three or four months after mating, depending again on temperature, the membranous 'eggs' are deposited one by one. Immediately the miniature adult inside ruptures the membrane with its head, struggles free, and slithers away to safety. From the word go, a baby adder is equipped with a full complement of fangs and poison, although it is rare – supposing it was born in late August or early September – for it to catch its first meal, probably a small mouse or shrew, until the following spring. It hibernates at the end of September or the beginning of October and, with the body just ticking over through the winter months, it requires very little energy.

Females tend to deposit their young close to the hibernation site, and as other snakes return to the same sites each autumn, large numbers are often found huddled together. This has the dual advantage of providing mutual warmth during the winter and concentrating the population for mating the following spring, when the animals will be rather low on energy. A few juveniles hibernate alone. Favourite sites, to the astonishment of telephone service engineers, are the conduits below man-hole covers.

Hibernation is not static. Movement takes place throughout the winter, albeit slowly, and on sunny days adders will emerge to bask, though the males always come out two weeks ahead of the females. Curiously, an adder kept in artificially uniform temperatures for an entire year will die. Periods of cooling, particularly seven weeks in the winter, appear to be necessary for the animals' biological clock to function and associated physiological changes to take place. In this way the body is ready to function in the spring irrespective of temperature. There is some evidence from work on the pituitary gland and hypothalamus part of the brain that the clock is set when the animal enters its first period of hibernation, and that the cooling period each winter retriggers the cycle, galvanizing into action other endocrine glands, such as the thyroid which initiates sloughing.

5
The Living World of Bees

About 100 million years ago, during the Cretaceous period, some wasps gave up meat-eating and turned their attentions to the newly emerging flowering plants, where they discovered the nutritious mix of pollen and nectar. They developed into a separate group of hymenopteran insects – the bees.

By far the greatest number of bees are solitary, the rest showing varying degrees of social organization. There are both social and solitary species of short-tongued mining bees, and long-tongued mason and leafcutter bees. There are stingless bees, and the highly social bumblebees and honey bees.

Bees prefer to live in warm dry places, such as southwestern North America and the lands bordering the Mediterranean. Many have been transplanted far from their original habitat by man – mason bees accompanied the slave traders from Africa to North America and the Caribbean; leafcutter bees started out in Europe and have now become one of the major pollinators in the USA; honey bees, also from the Mediterranean, have been taken to all parts of the world, whilst African varieties have been translocated to South America, with diasastrous consequences.

The Dance of the Bees

In 1967, the eminent Austrian zoologist, Karl von Frisch, published a remarkable account of the way in which forager honey bees, returning to the hive, tell others where to go to find the best sources of pollen and nectar. They perform a dance, and during that dance the bee communicates information about the direction of the source with respect to environmental clues – mainly, it is thought, the sun.

When a foraging worker returns to the hive she is 'frisked' at the entrance, for only bees that 'smell' correctly are allowed to enter. When she gets inside she begins to gyrate and run about in a peculiar manner. Such outlandish behaviour naturally attracts the attention of others in the colony and they gather around gently pommelling her with their antennae. Having captivated her audience the coryphee begins her performance.

On returning from a source of food in close proximity to the hive, she will simply perform the 'round dance' and move around in a circular pattern first one way and then the other. The vigour with which the dance is performed, and the rate of reversals, conveys information about the quality of the nearby food, but nothing about where it is to be found.

If, on the other hand, she has returned from a distant food source, she waggles across the comb at a particular angle, stops waggling, and then circles back to the right, waggles across again at the same angle, stops waggling and circles back to the left, and so on. The angle that the waggle part of the dance makes to the vertical represents the angle between the sun's azimuth and the food. The length of the waggle section and the number of waggles represents distance, and the intensity of the waggle indicates richness. All the while, the 'waggle dance' is accompanied by a buzzing sound that is perceived by the surrounding workers through their antennae and legs. This also gives information about distance and quality. The means of conveying information is all the more remarkable when you consider that the bee must convert horizontal information into gravity-related vertical movements, for the honeycombs, on which she performs, are set upright in the hive.

A food source directly below the position of the sun will elicit an upward-moving vertical waggle. Food located, say, 45° to the east of the sun will result in a dance that is 45° left of the vertical, and a source 180° away will be communicated in a vertical waggle down the comb.

Curiously, though, the bearings are not exactly matched. There is a slight misdirection, following a regular diurnal pattern, occurring in the dance. The angle of the gravity-assisted 'waggle' appears to be influenced and offset by the lines of the earth's magnetic field. Bees, it seems, can detect and appreciate the force of magnetism.

In the body of the bee, bands of cells containing magnetite associated with nerve fibres have been found concentrated around the nerve centre (ganglion) in each abdominal segment. From each ganglion a nerve branch enters the magnetic tissues. It has been suggested that the small magnetite particles twist in response to the earth's magnetic field and produce a torque which could be detected by the accompanying nerve cells.

When the sun goes in the bees have back-up systems to locate favoured food sites. On cloudy days, with patches of blue sky, or in forests where the sun would be behind foliage, foragers analyse the polarization of ultraviolet light from the clear part of the sky and work out a bearing on the position of the sun which they can then use in the dance. On overcast or foggy days, bees still continue to dance and collect food, and are able to predict the position of the sun with reference to local landmarks. And if there are no landmarks, they can still find distant sources by relying on their magnetic sense.

Househunting

After a period of intense brood rearing, usually in late spring, a colony of honey bees is ready to swarm. The nest becomes overcrowded, and the chemical messenger (pheromone) produced by the queen to control the colony and suppress the development of reproductive females becomes progressively weaker. New queens are reared and the old queen, together with half the workers, leaves the colony and looks for a new home.

The swarm flies just a short distance at first and gathers in a great mass, sometimes 15,000 insects, on the branch of a tree. Here they wait and rest as a few hundred scout bees fly off, as much as 10 kilometres (6 miles), to search for new premises. The reconnaissance party is made up from the older foragers that know the surrounding area well, and when they return with news of a suitable site, they perform a zigzag dance interspersed with bouts of wing buzzing through the dense swarm. Again, the line and the tempo of the dance indicate the direction and distance of the prospective nest.

Bees are fussy about the size and position of their new home. They need a cavity large enough to build sufficient honeycombs for winter storage, yet not too large that it cannot be kept warm when the weather gets colder. The entrance must be small and easily defendable, about 2m (6ft) from the ground to reduce the chances of disruption from ground predators, and face south in order that foragers can warm up on cloudy days. Entrance holes at the bottom of nests are preferable to those at the top to minimize heat loss. Cracks and crevices, that might produce drafts or let in rain, are not important in the selection process for they can be sealed with tree resins. Inspection takes about forty minutes.

With so many scouts and such a range of suitable nests, the bees must reach a consensus on which site to choose. Each scout returns with news of a possible site and performs its dance. Some dance more enthusiastic-ally than others, for it is the liveliness of the performance that indicates the degree of suitability of the site. When a more restrained dancer meets an enthusiastic one, the former obtains the information about the latter's site and heads off to look for itself. Eventually, all the scouts will have inspected the site of the liveliest dancer and the entire corps de ballet will be performing the same dance.

While the scouts are out and about, the rest of the swarm becomes inactive. Their body temperature drops to such an extent that if you shake the branch they will drop in a heap on to the ground; their wing muscles are too cold to work. Just before the flight, however, the swarm warms up to a temperature at which flight muscles work best, the links between bees break down, and the swarm moves off, one layer at a time, accompanied by loud buzzing. At first they circle above the resting site in a 10m (33ft) diameter cloud as the scout bees fly across the swarm indicating the direction in which they should all head. The swarm then departs, slowly at first, eventually streaking along at 10kph (6mph), skimming the top of the vegetation.

At the new nest site, the scouts drop down and release a special pheromone that causes the rest of the swarm to gather around. In a noisy whirlwind, the swarm streams into the new nest chamber and begins to clean out rubbish and construct new combs – the bees have a new home.

Undertakers

The closer we come to looking at the behaviour of bees the more suprises we find. In the honey bee hive, for example, bees carry out particular tasks at different stages of their working life. They help to feed and clean the larvae, build new honeycomb and larval cells, tend the queen, look after the eggs, ventilate the hive, search for new food sources, and gather the nectar and pollen. Now researchers have discovered a new occupation – undertaker.

It is important that large numbers of animals living in close proximity to one another, such as in a bee colony, keep the living space clean and thus minimize the risk of spreading disease. So, one to two per cent of the colony take on the role of undertaker. Surprisingly, corpses are not difficult to locate for when a bee dies it releases a special pheromone that can be detected by the undertakers. Within an hour of death a dead bee will be removed, grasped in the undertaker's mandibles, hauled to the entrance, and carried up to 120m (400ft) from the hive.

Giant Bee

Over a hundred years ago, the English biologist Alfred Russell Wallace was gathering information and specimens in Indonesia, that would one day lead him to the same conclusions about evolution as Darwin, when he discovered the world's largest bee, known simply to western scientists as the giant bee and to the local villagers as 'king bee'. After this encounter it was not heard of again and was thought to have become extinct. Then, in 1981, a young graduate student, Adam Messer, from the University of Georgia, rediscovered the bee living on an Indonesian island and has subsequently found it on three other islands, although the species is very rare.

The giant bee is enormous. The female is about three or four times the size of a honey bee, and is characterized by huge pincer-like mandibles. These are used to collect resin from wounds in trees. The bee scrapes together a blob of resin and carries it back to the nest, the ball firmly grasped in the mandibles. The nest itself is to be found inside the paper nests of tree-termites. The bees isolate themselves from the termites by building a tube of dried resin. They are not social bees, although several females have been seen to share a single nest. They do not cooperate and there is no division of labour. They have a barbless sting for defence. The males are about half the size of the females.

Polyester Packaging

Many species of bee nest underground and must keep their homes dry, warm and watertight to reduce bacterial and fungal diseases in developing larvae. Some bees prepare a silk lining for their burrows and chambers, others use a waxy secretion, but there is one genus of bee, *Colletes*, which makes polyester.

The discovery was made at the US Department of Agriculture's Beltsville Research Center, outside Washington DC, by Suzanne Batra. Previously, it was thought that the bee produced its peculiar secretion from salivary glands but after observing their underground activities from a glass-sided tank in a darkened chamber, Batra found that this was not the case.

The bee digs a burrow and then starts to tumble over. In little over a second it licks a substance from its sting and spreads it across the soil using its brush-like mouthparts. Using this tumbling and spreading action it lays down a thin film of waxy polyester, and by spiralling up the tube lines the sides of the burrow.

The polyester is formed in a similar way to that of man-made fibres. Two secretions with compounds consisting of carbon and hydrogen rings are produced from the Dufour's gland which opens at the tip of the abdomen of female bees. A stomach or salivary enzyme causes the rings to break apart and the compounds rejoin to make a solid lining. The polyester bag is not woven; it is simply a thin sheet.

Curiously, *Colletes* leaves a small loose flap towards the bottom of the cell, and it took many diligent hours of observations to see why. After a day constructing the burrow, the bee leaves for the night and returns in the morning. It climbs down, fills the chamber at the end with pollen and nectar, and lays an egg. The flap is then used to seal the chamber. The larva can then develop and grow, isolated as it is from the outside world. And all that effort pays off, for *Colletes* has the best survival rate of any of the burrowing bees.

DDT Bees

All the houses in the villages along the Ituxi River in Brazil are sprayed regularly with DDT in an attempt to control malarial mosquitoes, but the programme is suffering an unexpected upset, for living in the same area is a neotropical orchid bee that likes DDT. The male *Eufriesia purpurata* flies into the houses where it spends two to three hours scraping the insecticide from inner walls and storing it in pouches on its hind tibia before taking it back to the nest. And the bees come back, time and time again, obviously having suffered no ill effects.

Researchers from the University of Brazil examined the tissues of dead bees and found that they had as much as 2,039 micrograms each in their 25,000-microgram bodies. Normally, a level of just 6 micrograms is sufficient to knock down honey bees. It is thought that DDT is similar in structure to a substance that the bees collect naturally. They most likely convert it to a sex pheromone or a territorial marker, or maybe as a defence substance against other predatory insects.

The researchers spoke to the inhabitants of the houses and they confirmed that bees rarely entered their homes before the spraying operation began. Three-quarters of those interviewed did not appreciate the arrival of the bees, mainly from July to September, because they made such a noise. Fortunately the bee is stingless.

As for the anti-malarial programme, the bees, it seems, have not interfered too dramatically, although it has been estimated that during three months of intensive collecting, a wall of 12 square metres will be stripped completely of its DDT.

Yellow Rain

In the late 1970s, military scientists from the USA claimed that the Soviet Union was sponsoring chemical warfare in Laos and Kampuchea. Their evidence was in tiny yellow spots, found on rocks and leaves in the forest, which contained several dangerous toxins. The researchers at the Pentagon, however, had not checked out all likely sources of such a phenomenon – in particular, they had reckoned without bees.

It took a group of scientists, led by Matthew Meselson, of Harvard

University, to propose a hypothesis which suggested that 'yellow rain' was in reality the faeces of bees. They simply took electronmicrographs of yellow particles found on leaves and on the windscreen of an automobile in Massachusetts and compared them with the official pictures from Laos. They were a good match. There were also bee hairs found on samples both from Laos and Massachusetts.

Bee expert from Yale University, Thomas Seeley, who has spent many years in southeast Asia, confirmed that bees often store up faeces for weeks or even months before flying away from the hive and voiding their waste products. They may lose as much as 40 per cent of their body weight after a 'toilet' flight.

The toxins, it is thought, do not come from the bees but from fungi that grow on the faeces.

Killer Bees

As the teacher arrived at the school she was stung by a bee. She tried to slap it but the bee must have released some kind of alarm odour and the rest of the swarm in a nearby tree responded. Suddenly thousands of bees surrounded the woman. She tried to flee but could not run fast for she had an injured leg. She tripped and fell into the ditch, but managed to crawl out, her face and neck coated in bees. Still the bees kept coming, and passing cars became covered. People from nearby houses brought water but they were driven back by the swarm. Firemen arrived and they too were beaten back, every man covered in bees. They returned with smoke torches and eventually were able to get to the teacher. It was too late; she was already dead.

Not an extract from a science fiction novel, but a true story which took place in July 1975 at Aracaju, in Brazil. The bees were not a local variety but ones that had been imported for a biological experiment that went tragically wrong. It is reported that several hundred people and countless animals have been killed after the bee-keeping accident in 1957 when 26 notoriously ferocious African bees escaped from a genetics experiment at a research station near São Paulo. By mating with the more docile wild bees, their numbers have been rapidly increasing, and a wave of potentially dangerous insects has been on the move. Some have now reached the southern states of the United States.

Anthony Smith, the zoologist, writer and broadcaster, once told me about his encounter with African bees in the Matto Grosso, then the front line of the advancing swarms.

Someone had seen some honey bees not too far from our camp and we went along, in traditional Brazilian style to try and collect that honey. The nest was in the hollow of a small tree, and we could see the bees going in and out about 15 feet above the ground. The cautious ones among us began piling sticks to make a reasonable blaze. Smoke tends to make bees gorge themselves with honey and they are then less ready to attack. However, there was a Brazilian with us who pooh-poohed our preparation. Bees weren't worth all that trouble, he said. It's just a matter of chopping down the tree. So, he took our axe, and we swiftly took our leave. He hit that tree just once. The bees set about him, and he was soon in our company, a wiser and pained individual who, like us, was prepared now just to watch the fury around that tree. The Africans had arrived.

The scientist responsible for the experiment from which the original queens had escaped was Warwick Kerr. He was trying to improve the productivity of local honey bees, themselves a foreign race of honey bees from Italy, and introduced the African race of the honey bee, which is known not only for its ferocity, but also for its high production of honey. He took sensible precautions to prevent escape. On each hive, for example, he put a queen excluder. This allowed workers to enter and leave the hive but prevented the fatter queens from getting out and so the colonies were unable to swarm. The work was of interest to local bee keepers, and Kerr maintained an 'open house' so that they could see what was going on. Unfortunately, one day when Kerr was away from the forest apiary, an uninvited bee keeper strolled in and noticed the queen excluder grids. He also saw that the workers were having trouble squeezing through without loosing some of their pollen, so, being a diligent bee keeper, he took it upon himself to improve the efficiency of the hive and removed the grids. The hives, of course, were more than ready to swarm and within a couple of days most had taken off into the surrounding forest. Within a year they had colonized areas distant from the apiary, the rapidity of the invasion due, in part, to cross-breeding with the local mild-tempered bees. In the cross, the traits of the African strain seemed to predominate.

The African strain of honey bees differs in several ways from the European ancestral stock. It is smaller, for example, and if given European comb foundation, will chew up and rebuild it on a smaller pattern. Isolated as it is by the barrier of the Sahara, it is thought by some entomologists to be close to becoming a separate species. The ferocity of the bee is thought to stem from its long association with man. There is evidence from very early rock paintings that man was interested in the honey bee right from the beginning, and so the bee in the wild probably regards man as one of its natural enemies. And man is led to honey bee nests by a bird, the honeyguide. The bird is interested in the wax from the honey comb, but it needs an animal like man or its other close associate, the honey badger, to break open the nest. For such a relationship to have become established, it would have taken many thousands of years.

African honey bees are well known for their unpredictability, particularly when it is hot and dry. In these conditions they can attack without provocation. Imagine the surprise then of the Brazilian bee keepers who had been used to the docile Italian strains. Previously, they had been able to keep colonies in simple hives, such as packing cases, and remove the honey with considerable ease. When the African strain took over the packing case hives, and the keepers tried the same techniques, they were in serious trouble, and many were killed or badly stung.

The success of the African invasion was due to its ability to take over established colonies. The African drones, for instance, are more successful in mating with the queens on the 'nuptial' flight than are the European drones. When the queens are about to take to the air, the drones mill about waiting for their arrival. It was once thought that the drones were from the queen's own colony, but it is now thought that the drones are a mixed bunch from many colonies. In the race for the queen the Africans come out on top.

The speed with which the African swarms moved across the country, in some cases about 320km (200 miles) a year, surprised many researchers; after all, the European honey bees only travel a mile or two each year during swarming. For the African bees in tropical South America there is no one swarming season. They can go at any time, all year round, and therefore might swarm every two or three months. The African colonies are also smaller, and so reach their point of swarming much earlier. They also seem to travel much further during each swarming event.

The new strain did not fan out equally in all directions. It preferred the savannah grasslands in the northeast of Brazil, an area similar to the homelands in Africa. It was thought, at first, that the bees might have been stopped by the vast tropical forest areas of Amazonia, where the rainfall is heavy. But they were not halted. They passed right through, leaving colonies in the forest, albeit at lower densities than on the grasslands.

Of most concern at present is the likely invasion of the USA. All places with at least 240 frost-free days a year are susceptible to colonization. The worry is that the Africanized bees prefer to collect from tree flowers and so vegetable crops such as tomatoes would be neglected. Bee keepers are also concerned that an increase in the number of people attacked might result in more of them being driven out of business when victims sue and win high compensation claims.

Two male adders in a ritualised fight (*Lionel Kelleway*)

A 'knot' of male adders (*Lionel Kelleway*)

Part of the White Cattle of Chillingham herd (*Stephen Hall*)

White Cattle — calf, bull and cow (*Stephen Hall*)

Worker honey bee collecting nectar (*Ian Redmond*)

Lion at a kill in the Ngorongoro Crater (*Ian Redmond*)

Lioness yawns in the Parc des Virungas in eastern Zaire (*Ian Redmond*)

When the Africanized bees reached the isthmus of Panama, in 1982, the US authorities decided to carry out some tests in order to see what they could expect to encounter. In one experiment, they compared the way European bees and African bees attack a target, consisting simply of a suede ball supposedly the size of a bee keeper's head, as it approached their hive. The killer bees responded immediately, long before the ball came close to the colony, and swarmed over it. They delivered eight times as many stings as the European control group.

In 1986, the bees reached Mexico in large numbers and there have been isolated swarms spotted in the USA. In July 1985 an oil-field worker was reported to have seen a rabbit being killed by a swarm of bees in the desert to the north of Los Angeles. The movement of oil-field equipment in containers from South America is thought to be a likely means of import. As a precaution, every wild bee colony within a radius of 80km (50miles) was eradicated. The main bee invasion is due in Texas in 1988.

Vulture Bees

A scientist working at the Smithsonian Tropical Research Institute in Panama put out the carcass of his Thanksgiving turkey for the local cats and found that it attracted an unexpected scavenger – a swarm of bees.

David Roubik had been studying bees, particularly the so-called 'killer bees', in the rain forest when he chanced upon a colony of stingless bees of the species *Trigona hypogea* and brought them to the research compound. The bees had often been seen hovering over dead bodies and it was thought that they were simply after the juices.

Closer examination of their mouthparts, however, has revealed that they have five large and pointed teeth on each mandible with which they can cut into flesh. Some members of the swarm will descend on a carcass and stand around in a circle tearing at the skin. They make a small hole through which the rest can enter and methodically demolish the insides of the dead animal. As they chew they spread an enzyme over the meat and partially digest it before taking it back to the nest where it is regurgitated to others.

When a carcass is located, a bee will lay down a pheromone trail to the nest and thereby recruit further help. By arriving at a food source in large numbers *Trigona* displaces competitors, particularly flies which are chased away – an attribute shared with social mammalian predators that dominate food sites by aggression and sheer numbers. Curiously, one species of ant, *Crematogaster*, is not attacked and, in turn, does not harass the bees, although Roubik saw a polybiine wasp killed by ants. On one occasion ants had been at a carcass for two hours before the bees arrived, but they gave way to the bees and only returned at night. It is thought that there must be some sort of chemical interaction between the bees and ants which causes them to exercise a shift system – bees by day, ants by night.

The most common carrion is amphibian or reptilian; a dead frog can be reduced by 60–80 bees to its bare bones in three hours, but a thousand-strong swarm may tackle a dead monkey or anteater and strip it bare in a few days.

6
Africa

Africa is bisected by the Equator and much of the continent is hot. Straddling the Tropic of Cancer, and dominating north Africa, is the arid Sahara Desert, the largest desert in the world and a vast, waterless tract of land with few creatures evident during the day but many emerging at night. In the south, the Tropic of Capricorn crosses the coastal Namib Desert, which gets moisture from sea fogs and plays host to plants and animals that can absorb water directly from the air, and the drier Kalahari Desert, where the Bushmen live in harmony with the environment just as their ancestors did 20,000 years ago.

In the basin of the River Congo lie the equatorial rain and swamp forests, still to be fully explored and, perhaps, hiding large and mysterious creatures, known to the local peoples, but yet to be discovered by western scientists.

The Great Rift Valley is a geologically active zone in east Africa and the product of massive earth movements that are splitting the continent and giving birth to a new ocean. Here live the great herds and the predators that follow them. Some believe it to be the 'cradle of man'.

King of Beasts

The lion today is confined to the African continent and the Gir Forest in northwest India, but it was not always so. Prehistoric cave paintings from 15,000 years ago show that lions were common in Europe, for instance, and you might have encountered one in the Middle East up to the turn of the century. By the far the greatest numbers in the wild at present are living in southern and eastern parts of Africa, and information about their behaviour and everyday life comes mainly from a series of remarkable studies in the Serengeti of Tanzania by George Schaller, Brian Bertram, and David and Jeanette Bygott, Craig Packer and Anne Pusey.

The basic unit of lion family life is the pride, which may consist of up to twenty animals – a dozen adult females with their cubs, and half-a-dozen unrelated mature males. The pride does not always travel together, for it can break up into smaller three-strong sub-groups. This happens when food is scarce or when prey animals are small. If food is abundant, then lions prefer to be in larger groups – they are less likely to have their meal stolen by hyenas, and can more effectively muscle-in on the kills of other carnivores.

The pride lives in its own home range, maximum size 400 sq km (155sq miles), and may overlap with its neighbours. The males maintain territorial rights by roaring and scent marking with urine. The males also defend their females against intruding males, and will kill any male who has ignored the warning signs. Occasionally, a group of marauding males will attack the resident males, drive them out, and take over the pride. Then a strange thing happens: they kill the cubs. By murdering the offspring, the females come into heat sooner than if the new males had waited for the cubs to be weaned.

There is no one dominant male; the one nearest to a receptive female retains temporary dominance during mating – an event lasting 20 seconds but which might happen 50 times during a day. The males within a pride do not fight. They must stick together in order to chase away outsiders and protect the rest of the pride. They rarely go hunting for, with the large mane, they are too conspicuous.

The hunt is conducted by the females. They identify a suitable target and fan out in an attempt to trap the victim in a 'pincer movement'. At a hidden signal they charge, sometimes reaching speeds of 58kph (36mph), and grab the prey, killing it with a bite to the throat or suffocating it by clamping on to the snout. One clever piece of behaviour during the chase is the way a running lioness will slap the hind-quarters of the prey animal and simply trip it up. Although they approach within 30m (100ft) with considerable stealth, the charge is rather haphazard and many swift-footed animals, able to run faster than lions, escape. About one in four lion charges is successful.

All the pride are allowed access to the carcass, although squabbles develop over small kills, and, when times are really hard, the cubs under 18 months can die of starvation.

Cubs are small at birth as gestation is relatively short for such a large mammal – a little over 100 days. The females, curiously, will suckle not only their own young but the cubs of other females in the pride. Youngsters take milk for their first six months, although they begin to eat meat from three months. At the age of 2½–3 years, the young males are kicked out to fend for themselves and they usually form small groups of related individuals that eventually contest established resident males for possession of a pride. At first, even though successful in acquiring a pride, they may not retain it for long – perhaps as little as a year. As they become more mature and experienced they may run a pride, beating off all challenges, for up to ten years.

Lions breed well in captivity, in fact too well and there is a glut. In the wild, though, it is a different story, for habitat destruction, hunting, poaching and agricultural encroachment has meant shrinkage in living space and a reduction in the numbers of prey animals.

The Bird, the Badger and the Bees

The honeyguide, a small African nest parasite related to the woodpecker, has a distinctive chatter with which it deliberately attracts the attention of passing animals. More often than not it is a ratel or honey badger that responds to the call, but sometimes mongooses, baboons and even local villagers are side-tracked into following the bird, which calls, flits away revealing a white patch below its fanned tail, and calls again, repeating the pattern until it has enticed the follower towards a bees' nest.

Most mammals go wild about honey, a factor that the honeyguide has twigged through goodness knows how many thousands of years of evolution, and they will set about dismantling the nest to obtain the sweet reward inside. The honeyguide is content to watch, patiently waiting for a tasty morsel to drop to the ground. But the bird is not interested in the honey, nor the bees themselves, but in the wax that makes up the honey-comb.

How the honeyguide digests wax is not clear, although some researchers have suggested that the process is aided by symbiotic bacteria in the gut. The bird can also obtain wax from sources other than bees' nests, such as from waxy scale insects. They also appear to be able to smell burning wax from some distance away, and will quickly home-in on a local bee keeper extracting honey from his combs. Indeed, so acute is their sense of smell that one Portuguese missionary in the sixteenth century reported how they flew to the church every time he lit beeswax candles at the altar.

More recently, the honeyguide has given up trying to attract people – supermarket sugar is more convenient to obtain than raiding a wild bees' nest. Enterprising youngsters, though, know how to growl like a honey badger and encourage the honeyguide to lead them to the bees.

Naked Mole-Rats

The mole-rats are rodents that have become adapted for a life underground, and the naked mole-rats are of particular interest for they have a social organization that is more like that of an insect than a mammal. There is a queen rat, which is attended by workers and also individuals that appear to be soldiers.

The adult naked mole-rat looks like a little wrinkled sausage with four huge incisor teeth at one end and a pink naked tail at the other. Their eyes are small and apparently functionless. They remain only to detect the currents of air that might reveal damage to their intricate tunnel system. Touch is an important sense in the burrow and the only hairs to remain on the naked mole-rat's body are those sensitive to touch.

Smell is also important. The queen exercises her control over the colony with the help of chemical messengers (pheromones). These she secretes in her urine, which is deposited in the colony's toilet and thus to the rest of her subjects. Close body contact within the nest also ensures that the chemical is well spread.

There are only two breeding adults in the colony – the queen, which lives in the main breeding chamber, and her consort. They have about a dozen young at a time, although a pair in a captive colony has been recorded as having 27 in one litter. The rest of the colony, although not sterile, do not breed but are divided into the non-working adults and the working adults. The non-workers stay close to the queen and give the appearance of being guards. The workers dig burrows, maintain the nest, and fetch food and bedding. There may be up to eighty animals in a single nest.

When a litter of young is due, the teats of all the non-breeding females enlarge. It is as if the colony had a communal pregnancy. It may, though, serve a useful purpose. The body of each worker is chemically alerted to respond to the arrival of the new generation and attend to the needs of the young, even though they are not their own offspring. The workers look after the young but do not suckle them; that is left solely to the breeding female. However, the workers' faeces form a supply of food for the youngsters that have been weaned.

If the queen is killed, several of the female workers begin to develop sexually. After a short time, and without fighting, one female is acknowledged as dominant and the others revert to their non-breeding status.

Burrow digging is a cooperative activity, with several workers forming a miniature production line. At the head of the column a worker digs out the soil with his huge incisors. Lip folds behind the teeth prevent the earth from getting in the mouth. The next in line gathers the loose particles behind its body and shuffles backwards. Those behind straddle their legs and move forward, each one changing to a soil-pusher when it reaches the front. Eventually a continuous chain is formed, those with their bellies against the floor pushing soil backwards and those with their backs against the roof straddling forwards. At the back of the line one individual kicks the soil outside the burrow, and with such force that a small mound, like an erupting volcano, is formed.

The burrows, used for foraging below ground, fan out from the main chamber, and can be part of a subterranean network as much as 1km (0.6 miles) in length. The animals collect the roots of plants and can burrow into the large underground storage organs, such as bulbs and tubers, hollowing them out while they are still growing in the ground.

Predators are few. There are snakes, such as the mole snake and the eastern beaked snake, that prey upon naked mole-rats, attracted, perhaps, by the smell of the colony. Raptors (birds of prey) sometimes detect the activity at burrow entrances and seize the unfortunate worker responsible for throwing out the soil.

Hammer-Headed Bat

A boat journey along certain West African rivers in the dry seasons – June to August and January to early March – may be interrupted by a most extraordinary sight. All along the river bank, for maybe 1.5km (0.9 miles), are rows of fruit bats making a horrendous noise and flapping their wings for all they are worth. This is a lek, attended by male hammer-headed bats who are exhibiting their prowess as suitable mates for the visiting females.

Events begin when the males leave their day roosting sites, fly to the lek, and, early in the season, jostle and fight for the prime positions in the line. Here they return at dusk each day, hanging from branches in two formal rows in the trees along the river bank. At first, there may only be a few animals present, but at the height of the season there may be as many as 150 fruit bats, each separated from its neighbour by a distance of about 12m (40ft).

The male bats, with their large hammer-shaped heads, are twice the size of the foxy-headed females. In order to make their loud call, the male's chest cavity is filled by a large bony larynx, the cheeks are pouched, the nasal cavities can be inflated, the mouth is funnel-shaped, and there is a grotesque 'nose-leaf'. The 'metallic' call is very loud and is emitted at between one and four times a second. The wings are flapped vigorously at about twice the rate of the calls.

The female arrives at the lek and flies up and down the row, like a shopper in a bazaar, hovering in front of some males as if she cannot decide whether to buy or not. Any male under scrutiny then pulls out all the stops and gives a loud 'staccato buzz'. The female may be impressed but she does not let on at first; instead, she continues to fly up and down, hovering in front of this male and that, until she returns to the same few males several times during the night. Eventually she selects one and perches upside-down alongside him. Often it is the same few noisy and well-placed individuals that are selected by all the females. Within 30 seconds mating is complete and the female flies away.

Elephant Caves

Mount Elgon was once an active volcano, probably the highest in Africa. But now it is dormant and much of its upper slopes have been eroded by wind and weather to leave a 10km (6 miles) diameter caldera surrounded by a ring of mountain peaks. In the valleys are numerous caves, and it is one of these that Ian Redmond visited, first in 1980 as a member of Operation Drake and again on a private expedition in 1981, to see a strange phenomenon.

'Imagine you are lying on a rock in total darkness. The night sounds of the forest are drowned by the splashing of a small stream cascading over the cave mouth, through the blackness comes a low rumbling. A deep reverberating roar echoes all around the cavern and rises to an unmistakable trumpeting. You are witness to one of Africa's least known spectacles – elephants underground.'

The cave is called Kitum, and it is to be found 2,400 metres (7,874ft) up on the slopes of Mount Elgon. At the wide, low entrance is a small waterfall. Inside, the undulating floor is strewn with rockfall and divided by a huge crevasse. There are fruit bats roosting in the cave, and several species of bird nest there, but for the herds of elephants that visit each night, the cave is an enormous 'salt lick'.

Heavy rainfall in the mountains means that the soil is leached of many of its minerals, and therefore essential salts are missing from the plants that form the bulk of the elephants' diet. In the cave are lumps of sodium sulphate that the elephants dig from the walls with their tusks. Frequent visitors to the cave are characterized by having short, stumpy tusks. Each time they dig in the cave the ivory is worn away. They gouge off pieces, pick them up with the trunk, and then crunch them with their huge molar teeth.

They find their way in the dark, moving slowly and deliberately, and must be alert for a large crevasse just to one side of the pathway at the back of the caves. The skeletons of two baby elephants were lying there when Redmond visited. On average they spend between four and five hours in the cave. They eat the salts for about a half-hour and then sleep. In the mountains the nights are cold and the temperature inside the caves remains at a comfortable 14°C (57°F).

The elephants are not the only visitors. Bushbuck, waterbuck, duiker, buffalo, monkeys and baboons also come to collect their ration of essential minerals.

One mystery is why the caves should be there at all. Redmond did not see any of the usual physical features which would indicate river or wave action. Might it be that the elephants have excavated the cave themselves? Redmond calculated that it would take about 100,000 years for elephants to dig a cave the size of Kitum.

'Given the age of Mount Elgon', suggests Ian Redmond, 'we are left with the intriguing conclusion that these caves are actually elephant salt mines, dug out of the cliffs by generations of tuskers.'

Pink Elephants

Elephants, like humans, take the occasional tipple and many elephants, like many humans, have a drinking problem. This is the conclusion of Ronald Seigel at the University of California at Los Angeles, who has discovered alcoholism in raccoons, goats, cows, and pigs. The alcohol, in the wild, comes from rotting and fermenting fruit and, after a particularly heavy session on the plum-like fruit known as mgongo, a herd of elephants is likely to go on the rampage and flatten a village.

A drunken elephant hollars a lot, flaps its ears continuously, bangs its trunk, shakes its head vigorously and leans against anything conveniently upright. Unsupported individuals sway about a lot or simply fall down and are quite incapable of getting up again until the alcoholic haze has dispersed. They are unsociable – teetotal elephants spend about 80 per cent of their time with a herd, whereas drunks are alone for more than half their waking lives.

They are, however, careful about the strength of the alcohol they imbibe. Seigel was allowed to experiment with some captive elephants in Californian game parks, and found that the drinkers always chose a hooch of about 7 per cent alcohol – the amount of alcohol in fermenting fruit. At each session they would polish off the equivalent of about twenty beers.

The reason that elephants acquire and cannot kick the habit, according to Seigel, is that, like humans, they want to forget and escape their everyday life and experiences. This was again tested by confining two elephants in a small compound and observing their behaviour. They were understandably stressed, drank three times more alcohol than normal, and fell over a lot. Interestingly, Seigel has found that over the past few years, with increasing poaching, continuous droughts, agricultural enroachment and habitat destruction in Africa, elephants are living in higher densities and are therefore under greater pressure and suffering from stress. Their answer, like that of many humans, is to get drunk.

Chimp Medicine

Chimpanzees have several herb plants that they use not for nutritional needs, but for medicinal purposes. The leaves of *Aspilia*, a relative of the sunflower, for example, are picked by foraging chimps, but instead of chewing the leaves, as they do with other plants, they simply roll them in the mouth and swallow, much like a human taking a medicine.

It is thought, in fact, that these intelligent primates are taking the plant to purge the intestinal tract of parasitic worms. Local villagers around the shores of Lake Tanganyika use *Aspilia* for cleaning infected sores, relieving stomach pains, and flushing out intestinal parasites. The leaves contain about five milligrams of a powerful antibiotic, Thiarubine A. North American Indians use plants containing the same compound to treat wounds and sores. Chimpanzees suffer from the same kind of diseases as humans so that it is quite conceivable that the chimpanzees are using these plants in the same way.

Gorillas

At 3,050m (10,000 feet) high in the damp and drizzly cloud-forests of the Virunga Mountains in Central Africa, there lives the most majestic and dignified of the great apes – the mountain gorilla. Together with its lowland cousin, who is to be found in the rain forests to the west, this animal is probably the closest living relative of man.

Several million years ago, the gorillas branched away from our own evolutionary line, but today they might give us some insights into the behaviour and ecology of our early ancestors – that is, if modern man will allow them to survive.

The mountain gorillas are rare – very rare. There are thought to be only about 350 individuals living in the wild. The western lowland gorillas, the ones you are more likely to see in the zoo, are more numerous. There are 9,000–10,000 in the impenetrable swamp forests of West Africa. The eastern lowland gorillas in eastern Zaire number about 4,000. The serious decline in the numbers of mountain gorillas is due, for the most part, to the reduction of their living space. The destruction of rain forest for agriculture has meant that this species in particular has been badly hit, and so today it is confined to the slopes of just six extinct volcanic mountains, where the surviving groups are closely observed by a dedicated team of gorilla watchers.

Natural historians, though, were slow to investigate gorillas. The animals were very shy of man and lived deep in the inaccessible jungles of the 'Dark Continent'. Several sixteenth-century and seventeenth-century explorers and traders brought back stories of large ape-like creatures, but it was not until 1847 that the first scientific report appeared when an American missionary, Thomas Savage, visited Gabon. He wrote: 'Soon after my arrival, they showed me a skull of a monkey-like animal, remarkable for its size, ferocity, and habits.' Having studied the skull, Savage realized at once that this was a new species of ape. He gave it the name *Gorilla*. 'Gorilla has an ancient origin deriving from an account in the 5th century BC by Hanno, a Carthaginian who had sailed along the west coast of Africa and had killed 'hairy people' – the gorillae.'

The French-American adventurer Paul de Cheioux was so impressed by Savage's accounts that he set out himself on an expedition in 1856 to Gabon. Many creatures became hunting trophies, including gorillas.

> Before us stood an immense male gorilla. When he saw our party he raised himself erect and looked us boldly in the face. He was a sight I think I shall never forget. Nearly six feet high, with immense body, huge chest and great muscular arms, with fiercely glaring large deep-grey eyes and a hellish expression of face which seemed to me like some nightmare vision. Thus stood before us this king of the African forest. Just as he began another of his roars we fired and killed him.

At about this time the gorilla began to gain an awesome reputation for aggressive behaviour. Paul de Cheioux's book, some believe, was 'doctored' by his publisher to make the creatures seem even more horrific. A few years later anatomist Richard Owen made matters worse by retelling stories that attributed some of the less desirable of human characteristics to gorillas: 'When stealing through the tropical forest, the local inhabitants sometimes become aware of the proximity of one of these frightfully formidable apes by the sudden disappearance of one of their companions who is hoisted up into the tree, uttering perhaps a short choking cry. In a few minutes he falls to the ground a strangled corpse.'

Gorillas were thought to abduct maidens, a myth perpetuated in movies like *King Kong*, but these were all fantasies. Rupert Gardner was one of the few naturalists who, at the end of the last century, carried out serious field work, and he did it from inside a protective cage: 'One day, as I sat alone, a young gorilla, perhaps five years old, came within six or seven yards of the cage and took a peep. He stood for a while almost erect with one hand holding onto a bough. His lower lip was relaxed and the end of his tongue could be seen between his parted lips. He did not evince either fear or anger but rather appeared to be amazed.'

Naturalists nowadays do not sit in cages but follow gorilla groups until they are accepted as part of the scenery and thereafter are able to study the animals in their natural environment with the minimum of interference. In the Virunga Mountains, George Schaller, of the New York Zoological Society, did the groundwork on the mountain gorilla, and six and a half years later in 1967 Dian Fossey was accepted by several different groups and she started a long-term study that is still running today, despite her untimely death in 1985 when she was murdered, it is thought, by poachers.

Gorillas, she found, were basically family animals living in stable groups. Each group is led by a dominant mature 'silverback' male who may be supported in the protection of the very cohesive family unit by a sexually immature male or a younger silverback. His harem consist of three or four breeding females and their offspring. The group may total about fifteen individuals.

The male can stand 1.8m (6ft) high and weight up to 160kg (350lb), most of it muscle. He has the breeding rights to all the females which he wins from other groups to build up his own family. He is also the family peace-keeper, settling squabbles between females by simply strutting between them or making a short series of staccato guttural sounds. 'I get the impression', Dian Fossey recalled, 'that all the animals are aware of his physical presence, and his body language is very powerful. Sometimes he gives just a simple look and there is no more squabbling at, say, a restricted feeding site where there is only a small plot of bamboo or a small clump of blackberries.'

Some of the gorilla groups under observation remained together for a long time. One group, identified as Group 5, which was studied from the start, occupied the same range and contained some of the same individuals many years later. It was led in 1967 by a silverback given the name Beethoven who was supported by two other silverbacks, Brahms and Bartok. The two youngsters left but Beethoven, nearly twenty years later, is still the leader of Group 5. Indeed, two of Beethoven's sons stayed with the group during some of that time in the hope, presumably, that they would take over when he died. One disappeared and the other is still waiting. As gorilla-watcher Ian Redmond once told me: 'It's a bit like "Dallas" or "Dynasty", you're always waiting for the next episode.'

Although the gorilla family is stable for long periods, from time to time individuals stray away. Young males, for instance, may set out to start their own families. At this stage they may not have developed the fine silver back and are known as 'blackbacks':

> When the blackback reaches sexual maturity, and there are no breeding opportunities in his group, he will leave to spend up to four years wandering the forest in search of a female. When a young female approaches sexual maturity and there is no chance of mating – maybe she is way down the hierarchy – then she will also leave the family, but she does not go it alone. She will be taken by an extraneous male. This also helps in outbreeding.

The young silverback's days in the wilderness and his attempts at procuring a female are fraught with danger. At first his approach to other family groups is going to result in him being chased away, but gradually, as his experience and confidence increases his displays – 'whoop' calls, chest beating, tearing of branches, and bluff charges through the undergrowth – become more effective and he might successfully challenge an older harem leader and entice away one of his younger females. Occasionally, a young silverback will attempt to take over an entire family by fighting the leader. One of Fossey's study gorillas, Tiger, was one lone gorilla that got into hot water: 'Tiger challenged the leader of Group 6, who was a powerful animal. In the fight Tiger lost a right upper canine tooth, and a couple of incisors, and his face was mutilated. It took a long while for his wounds to heal and for him to try a challenge again.'

If Tiger had succeeded and had become the new group leader then it is likely that he would have systematically killed all the small babies in the family. By destroying the most recent offspring of his defeated rival he reduces the time it takes for the females to return to oestrus and thereby be available to have his offspring. A female might even gain status by mating with an infanticidal male who has killed her baby.

The gestation period for gorillas, like humans, is about nine months, and births can occur at any time of the year. At birth, the baby gorilla is pink, the black pigment not developing until later. It is about half the weight of a human baby and is totally dependent on its mother. Although it has a grasp reflex like the human baby, the gorilla is unable to hold on to its mother's fur and so must be continually supported. The mother then walks on 'all threes' instead of on 'all fours', one arm being used to hold the baby to the breast. Like the human baby, the neck muscles are not able to support the head, so that when suckling the mother will support the head in the crook of her arm.

Intervals between babies seem to be about four years. This seems to have some link with the way that gorilla mothers suckle 'on demand'. The stimulus of the baby at the nipple seems to inhibit the release of more eggs and so the female is not ready to have further young until the current one is weaned.

After about four to five months the youngster begins to make little excursions away from its mother. As it is beginning to get too big to be carried ventrally (that is, underneath) it tends to travel on her shoulder or back, and it will do this for about two years.

The living space in which the gorilla family leads a nomadic existence can be 5–30sq. km (2–12sq miles). Although it may be exclusive to one group there is no indication that it is a firm territory, simply a home range with a core area which they frequent the most. Neighbouring ranges tend to overlap but it is seldom that a group will penetrate the core area of another group. Spacing is most likely maintained by sound. The chest beats and 'hoots' of the silverback can travel considerable distances.

Communication within a group is probably more complex than we first imagined. At present we can only hear snorts, grunts and whines and their meaning is as yet obscure. One noise that is constantly heard when a group is foraging and feeding sounds a little like someone clearing their throat. This seems to mean, 'I am here, and everything is OK.' When feeding in the forest, many of the animals in a group are out of visual contact and this low 'cough' probably reassures the group members of the others' whereabouts. If individuals are particularly happy they appear to 'sing' with a high-pitched whine. A roar or a bark is used to alert the group to approaching danger and warn off any intruders. The chest beat seems to express tension as a result of fear or excitement. When playing, even youngsters will beat their chests as they chase about in the foliage, and immature blackbacks will strut about banging their chests to impress females.

Other means of communication are also used. The dominant male's silver back is a clear visual signal to the rest that he is the boss. Smell is also important. Silverbacks have a different sweaty smell, for instance, than the others, and this is particularly noticeable when the animal is displaying or sexually excited. And there is touch. Gorillas embrace each other a lot. If one group member is worried about the intentions of another then it is likely that he or she will simply reach out and touch the other. Ian Redmond remembered an occasion when Beethoven approached one of his females who had just had a baby, in fact her last before she died of old age:

He embraced her face to face with the baby between them. He was making the comforting rumbling sound and she was responding with little whimpers that seemed to convey affection, if a scientist dare use that word. And, as Beethoven moved away from the embrace she just left a hand on his back and watched him walk away. It was a very touching scene for a human observer.

Gorillas are essentially vegetarians. They like leaves, fruit and shoots, particularly goose grass, wild celeries, stinging nettles, thistles and bamboo shoots. They forage by day and rest at night. The gorilla day starts at about six-thirty in the morning, although on bad weather days they have been known to 'lie-in' until nine-thirty. They forage for about three hours, or until they are amply filled, and then rest. In the early afternoon they start to feed again, unless it is a sunny day when the siesta is prolonged, and then feed until dusk. When it begins to get dark they build themselves a nest. This can be a simple platform in the fork of a tree or a more complicated 'hammock' made of vines.

Gorillas have few enemies in the forest. Leopards have been known to attack but a silverback is more than a match for a big cat of that size. Humans are by far the greatest threat, for we destroy the habitat and hunt the gorillas.

Ironically, the very rare mountain gorilla lives in one of Africa's oldest national parks, yet the park and its gorillas have been under constant threat. The Mountain Gorilla Project run by a consortium of wildlife charities is helping to combat one of the major problems – poaching. Roger Wilson has been the field officer in Rwanda, and he has found that the gorilla deaths have often been a byproduct of other poaching activities:

Most poaching in the park is for antelope and for these they use the snare. It consists of a loop of string attached to a bamboo pole which is bent over. When the animal places its foot into the loop the bamboo is sprung and the trap tightens. Unfortunately, the antelope traps also catch gorillas. Young gorillas see these traps and go off to investigate and they get caught themselves.

The gorillas are not able to remove the snares and the wire or string may bite into the flesh. The wound becomes infected, goes gangrenous, and eventually, in a great deal of pain, the gorilla dies. Some of the gorilla deaths, though, are not accidental. Roger Wilson says:

A person who is going out to catch gorillas is after two things – the heads of the silverbacks, and often the hands too, or a live infant. For this they will use a gun, or more commonly a lance. The idea is to trace the group and get as close as they can. The male charges and they kill him with the spears. This is not as difficult as it sounds as a gorilla will charge the person who is annoying the group and all the poacher has to do is stick his spear in the way. The gorilla will run onto it. I've had one case where we had a silverback with a single spear wound running from the neck right through the abdomen. It must have run onto the spear at a crouch. In this case the poachers were after an infant. They had killed the silverback and injured the mother but had not, as usually happens, killed her. They had, though, taken the baby. Then, the group wandered around without a leader for several weeks and eventually amalgamated with a second group. So already in that incident we've lost two animals, and a whole family, with its home range in the park, has been suppressed. When the group joined up with the other group, there was infanticide, so there was one more death. When the park wardens caught up with the poachers, they threw down a sack. Inside was a baby gorilla which was so small it could fit inside a beret. We managed to keep it alive for three months, but it got dysentery and died. We did our best but the taking of live young is the most destructive form of poaching.

Physical damage, though, might not be the end of the story because gorillas, particularly the youngsters, have a strong emotional attachment to others in the family. A distressing series of events took place when poachers killed several members of Group 4. Ian Redmond recalled:

One individual was injured but not seriously. The wound would have healed, for gorillas have remarkable powers of recovery. What was more important was that both his parents had been killed and without a parent to give him the emotional security and stability the youngster's condition gradually declined until, three months later, he also died. His social world had been shattered, and that had an effect on his physical health.

Despite the efforts of the guards and wardens, however, as I write poachers are working the forests of Rwanda and Zaire to satisfy an inexplicable desire on the part of some people to possess macabre trophies – a silverback head, stuffed and mounted on the living room wall, or an ashtray made from a gorilla's hand. The brutality so vividly documented by those early naturalists is clearly misplaced; it is *we* who fit those descriptions, not the gorilla.

7

The Living World of Ants and Termites

Ants and termites have similar lifestyles – they are both social insects with a caste system of work delegation living in large colonies – but they are not at all related. Termites, confusingly also known as 'white ants', are related to the cockroaches, while the ants are grouped with the bees, wasps and saw-flies. Both ants and termites are widely distributed throughout the world, although termites tend to confine themselves to the tropics.

Termites are vegetarians, whereas ants take both plant and animal food. Indeed, ants have diversified considerably – there are slave-raiders, leaf-cutters, seed-crushers, harvesters, weavers, predatory armies, aphid farmers, those that cultivate fungus gardens, and others that collect nectar and store it in the crops of individuals that look like enormous 'honeypots'.

Slave-Raiders

Ants live in colonies, with egg-laying queens and sterile female workers. For most of the time, eggs hatch into larvae that develop into workers. At breeding time, males and new queens are produced, and mating takes place when the young queens leave the safety of the nest to find a partner during the nuptial flight. After copulation, the male dies and his sperm is stored in a special sac in the queen's body where it can be used to fertilize eggs for many years. With mating finished, the queen flies to the ground, rubs off her wings, lays her first eggs and raises the first few workers of the new colony. In over 8,000 species of ant, the workers tend the eggs, feed the larvae, build and extend the nest, search for food, and in some species cultivate it too. In just 30 species, the workers do none of these things. Instead, they enlist the help of other species by taking slaves.

A young slave-keeper queen, after mating, immediately takes over the nest of the slave species by chasing out the queen and workers but retaining the larvae and pupae which will become the first generation of slaves. They look after the queen's first batch of eggs, destined to become the slave-keeping workers. These workers do not carry out any of the colony's chores but scour the countryside in search of slaves.

Howard Topoff, a psychologist at Hunter College in New York, has been studying the behaviour of the red, western slave-making ant *Polyergus breviceps* which lives in the Chiricahua Mountains of southeastern Arizona, and has observed the events that take place when the colony leaves on a slave-making expedition.

For the best part of the day, the nest is quiet, with the slave-workers, usually *Formica gnava*, going about their everyday functions – tending the queen and her eggs, feeding the larvae and the slave-keeping workers, building and guarding the nest, and foraging for food. At about three o'clock in the afternoon, up to a thousand slave-keepers begin to muster at the nest entrance and, after a short while, a half-a-dozen of the oldest and most experienced workers set out to search for potential slave nests. When a scout finds a nest she heads back home and passes on the information to as many other workers with which she is able to touch antennae. The word gets around fast and at a hidden signal, the ants, both those near the entrance and others waiting inside the nest itself, set out towards the target.

The column, led by the scout ant, may be a metre wide and five metres long and it may gather still more individuals on the way. The scout finds her way back to the vicinity of the target nest using the position of the sun, while the rest of the marauders initially orientate to the sun but get their detailed directions from the scent trail laid down by the scout. Topoff was able to confuse the column, using a tarpaulin and mirror to shift the apparent position of the sun, and send the ants in the wrong direction.

Not far from the target, the scout and its followers suddenly stop. It is as if the scout has forgotten the exact location. So, a batch of the front-line workers fans out in an ever-widening circle, looking in nooks and crannies, under stones and under leaves, in order to find the nest. When an ant hits the jackpot all the others swarm towards it and the attack begins.

The raiders do not kill the other ants but simply spray them with a chemical – known as a propaganda pheromone – which causes them to run from the nest. The defenders attempt to carry away as many of their pupae as they are able but many are left behind. The slave-makers capture upwards of 3,000 pupae, and carry them, navigating by the sun and following the scent trail, back to their own nest. There they hand them over to the slave-workers and a new generation of *Formica* slaves is brought up to serve their *Polyergus* masters.

When new slave-making colonies are to be formed, the newly mated young queen does not head out to find her own site but returns to her home base. She then tags on to a raiding party and remains in the target nest after the others have left. When the residents return they do not attack this intruder but adopt her as their queen. They look after her first batch of eggs, destined to be the first slave-keeping workers, and a new slave colony is established.

Life amongst the slave-keepers is not always peaceful for there is considerable competition between nests, evident when the raiding party from one colony meets another at a territorial boundary. Often a battle ensues, and the stronger colony will take over the brood of slaves of the weaker. Unlike the slave raids, the battles are bloody, with ant killing ant. *Polyergus* workers have large and powerful mandibles, well capable of drawing blood from a human finger and certainly capable of slicing an opponent in two.

Studies of other slave-making ants, such as *Harpagoxemus* species, have shown that hierarchies develop amongst the slave-keeping workers. Nigel Franks at Bath University and Edward Scovell of Harvard University, have watched the behaviour of *H. americanus*, a slave-making species that lives in very small colonies, consisting of a queen, twenty slave-keeping workers and about 200 slaves. Encounters between workers sometimes result in ritualized fighting, while at other times a worker simply flees from another. Dominant workers pommel the heads of subordinate ones with their antennae and climb up on their backs, pinning them to the ground for a few minutes. The advantage in being at the top of the hierarchy is apparent at meal times. The dominant workers demand and receive more food from the slaves, and even muscle-in on the feeding of subordinates, driving them away in order to be fed by their slaves too.

The subordinate workers are charged with the duty of scouting for raids, and occasionally they take their chance to better their lot by remaining behind at the invaded nest. They then take over as slave-keeper when the host workers return.

An Army on the Move

Army ants are voracious predators. Great columns of ants on the move are capable of killing every living thing in their path. A colony might contain up to half a million individuals.

During the day the queen, her attendants, and workers carrying cocoons, make camp in a safe and sheltered place. Columns of ants radiate from the base with busy workers scurrying along on the inside and large, pincer-jawed soldiers guarding the flanks. Each file is on the look-out for food. Anything that cannot escape in time is taken, no matter what the size. A tethered horse or cow, or even an injured person can be stripped to the bone in a few hours. In one report from Goianira in central Brazil a column of ants 1.6km (1 mile) long and 800m (½ mile) wide was seen advancing on the town. The police chief and a few of the townsfolk were reputedly killed and the advance was only halted by men with flamethrowers.

The improbable tales about army ants and their African cousins, the driver ants, are legion but not all are horror stories. The inhabitants of South American villages are said to welcome an invasion of ants for they remove all the pests from the houses.

At night, the entire colony moves away, but these diurnal movements are not constant throughout the year. At certain times, determined by the readiness of the queen to lay eggs, the colony remains at one site, and for about three weeks the queen, surrounded and protected by her subjects, deposits over 100,000 eggs. At the same time the previous generation, which has been carried about in their cocoons by the workers, hatch out to take their places in the ranks. With egg-laying and hatching complete the colony once more returns to its nomadic way of life.

The way in which the colony lives and works in harmony is remarkable. Confronted by a stream, for instance, some workers interlock their legs and form a living bridge so the rest of the column may pass safely over. If caught in a flood, they all roll into a huge ball and float to safety, the individuals immersed in the water sacrificing themselves for the survival of the colony as a whole.

Accompanying the marching columns are often ant birds. These clever little passerines have discovered that an ant colony on the move will flush other insects out into the open where they can be picked off with ease. And the procession does not end there. Following the ant birds are ant butterflies. These ithomiine butterflies feed on the droppings of the ant birds and locate the ant columns by detecting odours given off by the ants themselves. The butterflies are protected from being taken by the birds by making themselves poisonous to eat. The butterfly larvae acquire poisons from the *Solanum* plant on which they feed and grow.

Weavers

Entomologists consider the weaver ants to be the most successful and abundant of the tropical social insects. They are also among the most ancient. Individuals identical to those living today have been found imbedded in amber 30 million years old. They live, for the most part, in trees where they build nests containing upwards of half a million ants. By ant standards they are large insects – up to 8mm (0.3in.) in body length, and they are remarkable for the way they construct their nests.

Nest sites are found when workers explore the territory, pulling on leaves to test for pliability. If an individual manages to grasp a leaf-edge and bend it round, others will rush over and begin tugging too. Very quickly there is a row of ants all pulling together. They may fold a leaf over or pull two leaves together. If the distance is too great for one ant to bridge then the others will form a chain, each grasping the one in front around the 'waist', until they can begin to bring the edges together. Other living chains form up alongside to help with the effort. With careful coordinated manoeuvring the leaf edges are gradually pulled together. At this point other workers appear holding larvae in their jaws, and the larvae are encouraged to exude silk (in a way akin to squeezing out toothpaste), to bind the leaves together. As the workers run back and forth with the larvae it looks a little like a cotton worker with a shuttle – hence the name weaver ants. The many strands of silk are woven into sheets and the tent-like nest is slowly built up. In some species, an entire wall of the nest may be made of silk instead of leaves.

Bert Holldobler and Edward Wilson, of Harvard University, have been studying weaver ants, and they have found that it is only the third instar larva that is used as a 'living shuttle'. It announces its availability for weaving probably by a chemical signal. The stages through which the worker must go in order for the larva to produce silk are very exact, and any deviation from the norm will hold up the supply. When a worker picks up a larva, it bends into the shape of an 'S', and when placed against the leaf and pommelled by the worker's antennae it produces silk threads from its labial glands. The glands are far larger than those of other ant species that produce individual cocoons for pupation, and externally the gland openings are considerably modified for nest weaving into a single long nozzle instead of the usual multiple openings. With this adaptation the larvae produce a single broad and tough silk thread suitable for the rigours of nest building.

The larvae, incidently, are not robbed of their silk for they are still able to produce their own cocoons a little later, when the nest is complete. The time at which they produce silk is simply brought forward, and this for the benefit of the entire nest.

Tree Defence

Wood ants climb trees and collect an incredible assemblage of tree-destroying insects. John Whittaker and his colleagues, of Lancaster

University, found that a single nest of a quarter of a million wood ants, entering and leaving the ant hill at a rate of about 16,000 per hour, captured and returned to their nest, during a period of one hour, over 100,000 harmful aphids and 2,000 destructive caterpillars. The research team was able to monitor this by intercepting the ants on their long trek home with electronic counters linked to pit-fall traps and gates around the ants' nest. They figured that this might be a way of assessing the damage that these insects might be causing the tree.

To test the hypothesis, they devised a way of excluding ants from trees so that they could compare the results with trees attended by ants. To do this they placed grease exclusion bands around the trunks of certain trees. The ants would not pass over the bands and could not, therefore, collect insects from the canopy. Then the researchers compared the difference in growth rates between the trees visited by ants and those without ants.

The conclusion was that trees that do not have their life-sap and leaf area forcibly removed by the myriad of insects to which the tree plays unwilling host can increase their growth rate by at least 30 per cent. It is thought that by carefully placing colonies of wood ants in tree plantations a degree of biological control can be maintained that will allow trees to grow to their optimum.

Ants and Plants

At the point where the leaf-like pinna joins the main stem of a bracken frond there are glands that secrete a sweet-tasting fluid. They are not active all summer, but just in the late spring and early summer when the new fronds are unfurling and the plant is most vulnerable to attack from plant-eating insects. Later in the season, when the bracken manufactures an array of toxic defence chemicals, the production of nectar ceases.

By producing the sweet liquid, the plant not only attracts flies and beetles but also ants, and ants provide the protection, by carrying off any insect about to feed and taking it back to the nest, during those early weeks. The strategy is not 100 per cent effective, in that some insects have developed their own formidable chemical defences against ants, and on plants guarded by ants they are able to proliferate in the absence of competition from other insect herbivores.

Similar relationships have been found with other plants, such as the flamboyant-flowered passion vine, found in fallow fields and along the sides of roads in the southeastern part of the United States. In addition to the nectar produced in the flower, at the bases of the bracts and petals, there is more secreted by a pair of glands situated on each leaf petiole.

These leaf nectaries are visited primarily by ants which have been recruited to protect the plant from insect grazers by carrying away caterpillars, grasshoppers and other grazers.

The nectaries in the flowers produce a limited supply of nectar at fixed times in the day when pollinating insects are likely to be on the wing, whereas the leaf glands provide a constant supply. The consequence of this is that five different ant species guard the plant, each one coming at a different time of day.

The aspen sunflower, which blooms in open subalpine meadows and at the edge of aspen forests in the Colorado Rockies, is prone to attack from tephritid or picture-wing flies. Adult flies settle on the plant to mate, after which the female lays her eggs on the buds or early flowerheads. The larvae munch their way through the ovaries and developing seeds and, in a good year for flies and a bad year for sunflowers, the entire seed production for the year can be gobbled up. The plant, by producing nectar from glands outside the flower, entices ants to visit the plant and take care of the pests. The ants are not actually very effective in catching the flies but they harass them sufficiently to deter egg-laying.

Bracts at the bases of the flowerheads of certain thistles also serve to attract ants that discourage flies from laying. Trumpet creepers have their extra-floral nectaries on the fruit, wild cherries in two bumps at the bases of leaves, and peonies on the buds.

In the tropics, many plants have built up an association with ants. The acacia tree, for instance, not only provides the food, but also the shelter. The ants live in the hollow thorns and feed on nectar from glands at the bases of leaves. In addition to keeping the herbivores in check, they have been known to snip away any creeper that threatens to smother the tree.

In Canada, forestry researchers have imported a particularly aggressive ant to protect conifer nurseries, and have found that the presence of ants reduces plant damage caused by insects. The idea is not new – in China, many hundreds of years ago, the citrus crops were protected by tree nesting ants and the farmers encouraged colonies to move from tree to tree by providing them with bamboo bridges.

Ants also provide a seed-dispersal service for some plants. There are, for example, about 1,500 species growing in Australia that depend entirely on ants. In one species of shrub, found in the arid scrubland of central Australia, the ants gather the fruit, take it back to the nest, eat their fill, and any left over is put on rubbish tips outside the nest. The tips are rich in nutrients and the seedlings grow far stronger than non-ant-dispersed plants.

Bone and Water Collectors

Harvester ants on the Athi Plains, not far from Nairobi in Kenya, collect the bones of small vertebrate animals. The relevation came when piles of unfossilized bones were found in rubbish heaps beside the entrances to nests. Outside one ant hill, for example, 1,167 pieces of mandible, teeth and other small bones from mice, shrews, small birds, and lizards were identified. It is likely that the ants are gaining additional protein by supplementing their normal grain diet with the small pieces of meat that would have been adhering to the bones.

Large black ants that live in desert conditions for five months of the year near Bangalore, in southern India, have developed a unique way of collecting water. The day is hot and dry but often a light dew forms in the early morning. Instead of rushing about collecting widely dispersed water droplets, they pile debris and dead ants outside the nest entrances where moisture can condense and be collected with ease. Even when there is no dew elsewhere, there is always some water on the rubbish piles and the ants spend the first quarter-of-an-hour of the day drinking their fill. Often as not, there is sufficient water for neighbouring ants to drink from the reservoirs without being attacked by the residents.

Architects and Civil Engineers

Termites are not related to the ants. They are social cockroaches and in some places just as much of a pest. Termite society is dominated by a large queen which is attended by workers and guarded by soldiers. There can be over ten million termites in a single nest but, in effect, the colony is like a single giant organism with each termite having its own part to play. They live in a fixed home base, known as the termitarium, which can be an irregular system of passageways and chambers in dry wood, or as complex as, as highly organized as, and equivalent in size to a modern skyscraper.

The simpler structures are built by the more primitive but much more damaging termite species. One type builds its galleries in the wooden beams of houses that suddenly collapse when load-bearing joists are hollowed out. So insidious is termite wood-boring activity that a wooden beam can be reduced to a paper-thin shell although outwardly looking intact. Ogden Nash saw some humour in it:

> Some primal termite knocked on wood
> And tasted it, and found it good.
> And that is why your Cousin May
> Fell through the parlour floor today.

In the French Antilles in 1809, British troops were able to capture a French fort with the minimum of effort. Termites had eaten away the garrison walls and when the French guns fired their first salvo the walls fell down.

At Isiolo in northern Kenya a colony of termites disrupted wedding celebrations. For safety, a farmer had buried a pile of bank notes in his garden. It was his life-savings and the dowry for his wife. The termites, however, reached the money first and destroyed the lot. The wedding was cancelled.

Fortunately these destructive termites are confined mainly to the tropics, although some species are found in temperate North America and one enterprising group has established itself in Hamburg, and is resisting all attempts to eradicate it. A block of apartments in Miami is being eaten away by Formosan termites that can chomp through wood six times as fast as the average termite and can apply their powerful acid secretion to destroy lead, plaster, asphalt, rubber and plastic.

In Africa, South America and Australia termite activity can determine the shape of the landscape and mounds may reach up to 10m (33ft) above the ground, the equivalent in human terms of a building 8km (5 miles) tall. The visible portion is usually constructed of chewed wood and soil particles mixed with saliva and faeces that can dry to the consistency of concrete. A termitarium can be a variety of shapes and sizes. *Macrotermes* builds enormous mounds with tall chimneys and arches. *Cubitermes* has mushroom-shaped roofs to shed heavy rains.

Inside a West African *Macrotermes* nest are interconnecting passage-ways and galleries with walls of softer chewed wood. Occupying a central royal chamber is the enormous sausage-shaped queen and her king. Her abdomen can be 14cms (5½in.) long and 4cms (1½in.) wide, for she is, in essence, an egg-laying factory – 33,000 eggs a day or one every couple of seconds. Because of her size the royal couple are incarcerated in their chamber for their entire life. A ring of soldier termites guards the queen and workers busy about her, bringing food, taking away eggs, and stroking and cleaning her rippling abdomen.

The eggs are placed in smaller nursery chambers surrounding the royal apartments. Here they are looked after by the workers. Nearby are larger galleries with convoluted walls and ceilings. In some, are placed piles of leaves and wood cuttings. In others, some of the colony's food is grown in fungus gardens. Leaf cuttings are mixed with termite faeces and the mycelia of a fungus that can break down lignin to make a substance more suitable for even the young termites to feed on. (The more primitive wood-eating species have bacteria or protozoans in a fermentation chamber in the gut. These break down cellulose in the chewed plant material and release carbohydrates which can be utilized by both bacterium and termite. The bacteria multiply so rapidly that the termite can cream off a few and supplement the protein in its diet.)

Radiating from the mound are tunnels in which the workers head out to forage for wood and leaves, and below the main structure deep channels, sometimes 40m (130ft) long, are dug down to specially constructed reservoirs at the water-table.

With several million tiny bodies, all busy, bustling and breathing and confined inside the termite mound, oxygen consumption is high. There can be a buildup of as much as 15 per cent carbon dioxide in the nest, enough to make a person unconscious. In fact, it was thought at one time that termites produced so much methane during their digestive processes that the outgassing could affect the world's climate. With such a high turnover of oxygen and carbon dioxide, termites have devised an ingenious air-conditioning system which is totally automatic.

Between the main gallery areas and the thick outer wall, there is a network of narrow ducts set in vertical ridges on the sides of the mound, and below and above the galleries there are large empty air-spaces. The air in the fungal chambers is warmed during the fermentation process and the air in the rest of the nest is heated by the metabolic processes of the termites themselves. This hot air rises into the upper space and is forced, by the continuous upward flow of warm air, into the narrow ducts at the sides of the mound. Along the rides, containing the ducts, the wall is thin and porous enough to allow an exchange of gases. Carbon dioxide diffuses out and oxygen filters in. The air coming in is cooler than that going out and so it flows down the ducts, via wider channels at the bottom, into the lower air-space or 'cellar'. The fresh air is then drawn up through the galleries to replace the rising warm air.

The East African termites have a slightly different method. Ugandan *Macrotermes* mounds do not have air ducts inside outer ridges but instead have flat chambers under the dome of the roof. Again, the walls are porous and the stale air can escape. The 'cellar' is linked directly to the outside by wide channels into which fresh cold air can enter the bottom of the nest. Another species of *Macrotermes* has tall, open-topped chimneys connected to ventilation shafts inside the nest. The shafts have thin walls through which the stale air from the galleries can diffuse, to leave the nest via the chimneys.

Termites prefer warmth and high humidity but they do not like it too hot. In Australia, the compass termites have an intriguing way of maintaining pleasant working conditions within the nest, despite it being exposed to the scorching heat of the midday sun. The mounds are huge, flattened slabs that stand with their broad sides facing east–west and their narrow end pointing north–south. This orientation means that the nest receives maximum exposure to the sun during the cooler parts of the day, at dawn and dusk, while presenting the smallest surface area during the hottest part at midday. But termites are, for the most part, blind – so how do they orientate their nests so accurately? Termites, like bees and a host of other animals, can detect and orient to the earth's magnetic field.

The termitarium is protected and insulated mainly by its concrete-like exterior, although some creatures such as the aardvark, the anteater, the echidna or spiny anteater and the marsupial numbat can break through with their sturdy front claws. Smaller predators, such as ground beetles and ants, which could easily enter the mound are dealt with by the soldier termites, a caste that comes in a variety of shapes and sizes with an astounding armoury of effective chemical and mechanical weapons.

Cryptotermes soldiers have a head shaped like a bath plug. When the mound is under attack they go to the narrow entrances and seal the holes with their heads, presenting an array of sharp mouthparts to discourage any intruder. *Anoplotermes* colonies do not have a soldier caste as such but each worker has a special muscle around its abdomen. If attacked it contracts the muscle and literally explodes in the aggressor's face covering it with acidic digestive secretions. It is the ultimate sacrifice, for in exploding the termite dies.

Macrotermes soldiers come in two sizes – large and small, and they are both sterile females. The smaller soldiers escort the big sterile male foragers outside the nest, while the larger ones guard the royal chamber. When ants invade, a soldier will use its powerful jaws in an attempt to chop off an intruder's legs. Heat, produced by the strenuous activity of the termite's muscles, melts a fatty secretion produced in the frontal gland on the termite's head. The oily substance flows on to its piercing mouthparts and is daubed on to the ant with every successful bite. The chemical enters the punctured exoskeleton and interferes with the ant's natural repair mechanisms. *Armitermes* soldiers produce a similar secretion that is toxic to ants, and *Rhinotermes* soldiers have a paint-brush-like structure on their head with which to daub contact poisons on to the cuticles of their attackers.

Soldiers of *Nasutitermes* species have chemical sprays. The head has developed into a pointed nozzle through which a defensive spray, secreted by a frontal gland, can be accurately squirted at an attacker. The substance is a viscous entangling agent that can quickly gum-up an ant and render it immobile. Chemicals in the spray irritate the ant, causing it to scratch and wipe itself, and so get further glued-up, blocking spiracles and sense organs. The spray also contains an alarm pheromone that is effective for 30mm (1in.) around the struggling insect. If one nasute termite locates and immobilizes an aggressor, soon many more defenders will gather to see it 'neutralized'.

One cunning creature has found a way of fooling nasute termite defenders. The nymph of the assassin bug *Salyavata variegata* of Costa Rica gains a meal with the help of a disguise and a bait – the first time an insect has been seen to use a 'tool'. The bug is able to approach the termite nest, usually located in a tree, by dressing up in bits of the outer covering of the nest which it attaches to its body with secretions of glandular hairs. Suitably disguised, it sidles up to an entrance hole, grabs a termite, injects digestibe enzymes, and sucks the insect dry. Soldier termites, coming to the rescue, crawl over and ignore the assassin for it feels and smells as if it were part of the fabric of the nest. When the bug has finished its meal, it uses the spent corpse as bait to capture another termite. It waggles the drained shell of the body in front of a tunnel entrance and inquisitive worker termites arrive to take it away. The protein in the dead termite's skin is valuable to the colony. The worker, lured into the open by the bait, is grabbed and eaten. The bug may continue, using the same 'angling' technique, for up to three hours in which it might devour thirty termites.

The ant *Decamorium velense* has evolved a similar chemical camouflage to outwit termite soldiers. Scout ants locate a termite nest and a strike force of about thirty break into the nest where they are able to move about quite freely, killing termites with their stings, and all without raising the alarm. It seems an alcohol secretion produced by this species of ant is effective in calming termites, whereas the smells produced by other ant species normally panic termites. Another clever little African ant, which cohabits with the termites, cadges a lift on the head of a soldier and grabs scraps of food as the worker comes to feed the soldier.

The termite guest rove beetle is allowed to live within the termitarium despite its habit of gobbling up young nymphs. Its white swollen abdomen is carried over its back and it produces a secretion that is attractive to termites.

Some larger creatures have learned to take advantage of the safe and warm conditions that exist inside a termite mound. In Australia, a tropical parrot, some species of kingfisher, and the lace monitor lizard excavate their own chambers in the outer walls of termitaria, and lay their eggs there.

In a termite colony there are several different types or castes of termite. The productive queen and king are the centre of attention and they are supported by aggressive soldiers, wingless sterile workers, and nymphs. There is usually a ratio of 100 workers to every soldier. The workers feed the queen, the king, and the soldiers. Unlike ants, bees and wasps, soldier and worker termites can be either male or female and both carry out work in the nest. There is no larval or pupal stage. The young hatch as nymphs and moult each time they need to grow a little larger. Their sexual development is arrested at an adolescent stage, leaving the queen and her consort the task of reproduction. Some workers, known as 'supplementary reproductives', are able to take over from the royal pair if they should be removed or killed. The entire operation is controlled by chemicals produced in the saliva and faeces of the queen and distributed throughout the nest by mouth to mouth contact between workers and soldiers. Occasionally winged termites capable of reproduction are reared, and these fly out of the mound in huge clouds to start new colonies elsewhere. The females from mounds that have fungus gardens always carry a piece of the fungus with them in order to grow a new culture in the new home.

Flying sexual termites have well-developed eyes. The male, though, locates the female by smell. Ants and bees have a once-only mating during the nuptial flight but termite kings and queens become partners for life. On reaching the ground, they lose their wings and the king, attracted by the queen's scent, runs after her in a 'courtship promenade' that may last for a couple of days. This is the time when they are the most vulnerable, falling prey to insect-eaters that appear to anticipate nuptial flights. The predators include man. Roasted termites are a delicacy rivalling, it is said, grilled prawns for flavour. Those few that survive find a suitable crack or cranny in which to start a new colony. The first chamber they build eventually becomes the royal chamber and the centre of the termitarium. Here they will spend the rest of their lives together for maybe ten or twenty years. When they die they will be replaced and the colony itself can continue to thrive and grow for centuries.

8
Australasia

Australia, New Zealand, New Guinea, the Moluccas, and Tasmania constitute the smallest of the continents – Australasia.

The fauna of Australia is, and has been, dominated by the marsupials or 'pouched' mammals. Discoveries of marsupial fossils in Australia, on the Antarctic continent and in South America have supported the concept of continental drift, plate tectonics, and the existence of the ancient super-continent of Gondwanaland. Australia is an island of diverse habitats – the scrub lands of the east, the tropical forests of Queensland, the sandy, stony, mountain, shield and clay deserts of central Australia (occupying two-thirds of the island), the luxuriant southwestern corner, the temperate southeast, and the Great Barrier Reef.

New Zealand, before the arrival of humans, was an island paradise for flightless birds and other relics of a predator-free environment – the hedgehog-like kiwi, the primitive tuatara lizard, the now-extinct 4m (12ft) tall moa, the kakapo, and the kea.

New Guinea (which some might assign to 'the East') shares many animals and plants with Australia, and is on the Australasian side of those great zoogeographical divides – Wallace's and Weber's Lines, which separate the fauna of the Orient from that of Australia.

Slow Cooking

When the European settlers began to explore Australia, they discovered curious mounds of earth in the arid *Eucalyptus* scrublands. At first they considered them to be aboriginal burial mounds, but the locals denied any involvement and insisted that they were made by birds. Nobody believed them, and it was not until 1840 that naturalist John Gilbert opened a mound and discovered a clutch of eggs. They were the eggs of what is now called the mallee fowl or lowan, a relative of the brush turkey, and one of a group of birds that does not incubate its eggs using body heat, but buries them in a heap of rotting vegetation and makes use of the heat produced during decomposition.

During the winter, before the rains come and the breeding season gets underway, a pair of birds will dig a large pit, about a 1m (3ft) deep and up to 5m (16ft) in diameter, and then scrape and carry every piece of dead vegetation within a radius of 50m (165ft) to fill it up. With the arrival of the rains, the rotting processes are accelerated and the temperature within the pit can rise to 60°C (140°F). The birds cover in most of the pit with a layer of sand about 50cm (20in.) thick, leaving a small section where they mix sand and leaves to form a chamber in which the female will lay her eggs.

The remarkable part of the whole story is the role played by the male. He must maintain the mound at a constant 34°C (93°F) throughout the period of incubation, and that can go on for many months, for the female lays her eggs, up to 35 in a season, in small batches over a long period of time. The male's is by no means a simple task, for, during the summer, the air temperature can fluctuate by over 10°C (50°F) during the day and night, and it varies considerably through the seasons.

In the spring, the male opens the mound in the early morning in order to let heat escape; in the summer he scrapes on sand to protect the eggs from the vicious heat of the mid-day sun; and in autumn, when the sun's rays are weakening and the rotting process is declining, he will spread out sand to warm it up before scraping it back on to the heap. The beak and tongue are inserted into the mound in order to keep a close check on temperature changes.

Development of the embryo starts immediately the eggs are laid, and the first chicks hatch out before the last eggs arrive. It might take them up to fifteen days to dig themselves out of the mound, but they enter the outside world fully feathered and able to regulate their body temperature. Within an hour, they can run rapidly and head for the safety of the scrub. They can fly after about 24 hours. They never see their parents.

The mallee fowl is not the only bird in its group to take advantage of free, environmental energy, but it is certainly the most sophisticated. The brush turkeys make mounds of rotting vegetation and control the internal temperature, to some degree, by removing leaves or putting them back. Other species simply bury their eggs in the warm sand at the top of tropical beaches, or utilize the heat in volcanically active areas or in the vicinity of hot springs. Forest birds may lay their eggs in rotting tree stumps or in piles of dead leaves and those living in rocky areas may deposit their eggs in a crevice exposed to the sun – the heat retention properties of the rock maintaining some of the warmth when the sun goes down.

Although the mallee fowl's method of egg incubation appears advanced, it is, in reality, based on a more primitive reptilian system. Many reptiles have underground nests, lengthy incubation periods at relatively low temperatures, and precocious young that receive no parental care after hatching.

A Bird in the Bower

Fifteen of the eighteen species of bowerbirds, living in Australia and New Guinea, build structures with which they hope to impress a female, as a prelude to courtship and mating. There are those that simply clear a 'court', others that put down a 'mat', and still others that construct elaborate 'maypoles' or erect 'avenues'. Often, the attractiveness of the bower is enhanced by brightly coloured decorations, the colour preference varying from species to species.

The tooth-billed catbird scrapes all debris from a large area on the forest floor and decorates it with fresh leaves, carefully placed with their undersides uppermost. Archbold's bowerbird goes one better and puts down a mat of mosses and ferns.

Simple 'maypoles' might consist of a single erect sapling surrounded by a conical-shaped column of interlaced twigs, and in the case of MacGregor's bowerbird, there is a platform of mosses and bearded lichens at the base. The platform consists of two concentric rings, one at the base of the 'maypole' and the other a few centimetres away, looking very much like a moat. The bower is decorated with all kinds of things coloured black, orange-brown or yellow. There might be pieces of charcoal, fungi, seeds, parts of insects, and lichens. On the column, the bird dangles insect frass, the silk-like material speckled with sawdust that is produced by wood-boring moths. Around the border of the bower, the bird places the leaves of screw pines, and in the foliage round about it hangs small bunches of fruits. If a decoration is moved, the bird will return it to its proper place immediately. No two bowers are alike, and this particular species might construct bowers anywhere between 30cm (12in.) and 3m (10ft) tall. The taller towers are the result of the bird coming back to the same site and adding a little bit extra each season. Some birds construct several smaller towers instead.

The most spectacular 'maypole' bower is erected by the Vogelkop gardener bowerbird of New Guinea. It consists of a 2.5m (8ft) long and 1.5m (5ft) tall, cone-shaped hut made of interwoven twigs supported by three or four saplings. At the entrance is a mat on which it places a multitude of decorative objects.

'Avenue' bowers consist of two parallel walls of twigs or grass stems standing on a platform of sticks. The walls are often painted, using a piece of bark, with charcoal moistened by saliva or the juices from squashed fruit. The walls of an 'avenue' bower of the satin bowerbird

Young lioness approaches others drinking at a riverside in the Serengeti...
(Ian Redmond)

...but appears to have taken the wrong place and starts a-scuffle
(Ian Redmond)

African elephants, in search of rock salt, file slowly over a mound of fallen roof inside Kitum Cave. The female at the rear keeps her trunk on the calf to prevent it from falling into a crevasse (*Ian Redmond*)

African elephant mining salt by gouging out the rock with its tusks and catching the pieces with its trunk before they fall to the ground (*Ian Redmond*)

Male bush buck in Kitum Cave (*Ian Redmond*)

Dead waterbuck in a crevasse (*Ian Redmond*)

Mountain gorilla country around Mt Mikeno in the Parc des Virungas in Zaire
(*Ian Redmond*)

Mountain gorillas — two juveniles and an infant (*Ian Redmond*)

almost meet at the top forming an arch. Decorative bits are predominantly blue or green.

In recent years, bowerbirds have been supplementing their natural decorations with man-made ones – car keys, bottletops, tooth brushes, spectacles, in fact anything that catches the eye is of the right colour. Satin bowerbirds have even been seen to paint their 'avenues' with blue washing powder granules.

The bower is not a nest site, but simply a place in which the male can entice and impress a female. Attention has been drawn to the fact that the birds with the less flamboyant plumage tend to build the most elaborate bowers. At every opportunity, a male will attempt to destroy his neighbour's bower or steal decorations. If, during an experiment, a bower is altered, moved or reorientated, the bowerbird will reconstruct it just as it was.

The satin bowerbird builds its bower so that the long axis is never more than 30 degrees off a north–south line. This, it is thought, ensures that neither sex has the sun in its eyes during the morning courtship display. During his performance the male tries to tempt the female into his bower. He picks up shiny objects, gyrates, bows and scrapes, and dances but, according to recent research, he rarely entices the female in and she flees without mating.

A female satin bowerbird is understandably nervous about a male's intentions, for mating itself is a violent affair. The male seems to attack the female, clawing and pecking her, often wrecking the bower and, indeed, sometimes chases her from the nest before copulation is achieved. When mating does eventually occur, the female crawls away exhausted, and builds her nest at least 200m (656ft) from the bower site. She brings up the brood without any help from the male.

Frogs in the Tummy

Many species of frog look after their tadpoles after spawning in order that the youngsters get a better start in life than their free-swimming cousins.

The tadpoles of the Australian pouched frog, for example, wriggle into pouches on the flanks of the male where they remain until they have developed into tiny froglets. There are two species of South American mouth-breeding frogs that look after twenty or so tadpoles in their vocal pouches. And the eggs and tadpoles of some species of Surinam toad develop in the safety of individual pockets on the back of the female. But the most bizarre brooding behaviour is, perhaps, shown by the gastric-brooding frog of the rainforests of Queensland, Australia.

As its name suggests, the female frog swallows about twenty eggs and they develop in the stomach. Somehow, the gastric juices are switched off so that the eggs are not digested, and several weeks later the tadpoles and froglets are 'vomited-up'. During the entire incubation period the frog does not eat.

Clearly, the implications of the research on this frog for those suffering from peptic ulcers and other stomach complaints was of the utmost importance, but this story is a sad one for not long after the frog's unusual behaviour was discovered, it became extinct. It lived in streams flowing through the Blackall and Conondale Ranges in southeastern Queensland and logging operations have been blamed for its demise. Fortunately, the Queensland National Parks and Wildlife Service saved the day for the researchers, and in 1981 discovered another species of frog that behaves in the same way. The research continued, and Mike Tyler and his colleagues of Adelaide University were able to reveal that it is a special chemical in the jelly surrounding the egg that switches off the mother frog's production of hydrochloric acid and other gastric juices, and also inhibits the peristaltic movements of the gut wall. When the tadpole emerges from the egg, it too produces the chemical, in strands of mucous produced from the mouth and in secretions from the skin. The chemical is one of a group of hormone-like substances known as prostaglandins. What stimulates the mother to regurgitate her offspring is not known.

The Booming Parrot

New Zealand was once a country without mammals and, with no major ground predators, many birds were able to live their lives on the ground, rarely taking to the air at all. Some became totally flightless. About a thousand years ago humans began to arrive, bringing domestic stock and releasing alien species as accidental or deliberate introductions – first the Maoris with dogs and rats and then the Europeans with cats, foxes and stoats. The new arrivals reduced many species of flightless bird to very small numbers. One bird, in particular, caught the imagination of conservationists for in the 1970s it was considered to be extinct. Then a small population was found in a remote corner of the Fiordland National Park – unfortunately, they were all males. A little later, fifty more were discovered in the southern part of Stewart Island – this time there were females present, and in 1981 they were seen to breed. Today, there are thought to be fewer than a hundred living in the wild. The bird is the kakapo.

The kakapo is the largest member of the parrot family, weighing up to 3kg (7lb), and covered with soft feathers streaked by green, yellow, brown and black. By day it lives in crevices or under exposed tree roots, coming out at night to feed on mosses, leaves, seeds and berries. Areas with lots of berries are known to the locals as 'kakapo gardens'.

The bird climbs trees in search of fruits and nectar, and it has rudimentary wings which not only help it maintain balance when rushing through the forest, but also allow it to launch off from a high bough and glide up to 90m (295ft) to another patch of vegetation.

The most interesting aspect of its life, however, is courtship. Kakapos, unlike any other parrot or bird in New Zealand, perform at a lek. At leks, male animals occupy small territories – usually the dominant ones at the centre of the arena – where they perform their courtship display. Females arrive, sample the performances, and settle for the best display. The chosen male mates with her and she leaves the area to rear her young elsewhere and on her own. The male stays put and attempts to attract further females. It is, in effect, an extreme form of polygamy.

At the top of a hill or small rise, the kakapo male digs several shallow pits in the ground. Some of these bowl-shaped hollows are excavated near structures in the environment that reflect sound well, such as a hollow log or concave rock face, and they are all linked together by well-trodden paths. At dusk, from late November to early January, the cock arrives at the bowl site and the display starts when it rocks to and fro and flaps its wings. After a while, the bird inflates its thorax to what has been described as 'gross proportions' and begins to give little grunts. These gradually get louder and louder until, every two seconds, they become 'sonic booms' that can be heard up to 5km (3 miles) away. It rests for half a minute, then starts all over again, and this performance can go on all night. In foggy weather the display has been known to last for 17 hours.

Until now, it was thought that animals display in a lek because there is safety in numbers at a time when they are exposed to, and may even attract, predators. The kakapo displays were evolved in an environment without serious predation, and so the scientists will have to reassess their interpretation of this kind of behaviour.

Sealion Summer

Enderby Island is a sub-Antarctic island far to the south of New Zealand. Each short summer, one of its beaches – Sandy Bay – is the site of an invasion by the rare Hooker's sealion.

At the start of the breeding season in October the males appear. They fight for the prime locations on the beach and by November, when the cows begin to arrive, each is lazily occupying its own territory. Some of the tawny-coloured cows haul out in the night and group together high on the dry part of the beach. Others arrive by day, recognizing the beach by sight and smell, and they emerge from the surf to join the others. But first they must run the gauntlet through the territories of the young, solitary bulls at the water's edge. These bulls have only the slimmest chance of keeping a cow, for those bulls *down* the beach are certainly way *down* the social ladder.

Curiously, the females determine where the breeding sites are to be. The mature bulls simply occupy territories in the most likely areas and the gradually increasing population of cows spreads across the territories of several dominant bulls. Usually there are about ten 'beachmasters' and although there is some degree of herding, the cows are able to move freely between the territories of several bulls.

The cows are ready to breed when four years old, and they return each year to the same place to pup and to mate. Before November is out, more than 400 cows will gather on Sandy Bay beach, many pregnant from the previous summer's matings and ready to give birth. Throughout December, the mothers move away from the crush of bodies and silently deliver their single pup. During the birth itself, the mother's pelvic bones move far apart, making the process both easy and quick. The pup, often still wrapped in its grey birth membrane, the amniotic sac, is dropped after about ten minutes, although labour can extend sometimes to 45 minutes. A new mother sniffs her pup frequently, and she is extremely aggressive, stoutly defending her pup against any intruders. For two to three days it will not leave her side, the pair learning to recognize each other by sight, sound and smell. It will stay with her constantly for the first year of its life.

Often the skuas, scavenging seabirds attracted to the sight of blood, descend on the breeding beach to share in the sealion births. They deal with the afterbirth, the still born, and the weak, but in a way they help the strong pups to survive by keeping the beach clear of decomposing flesh and perhaps reducing the spread of disease. Sometimes, though, over-eager skuas, in their haste to get at the afterbirth, damage emerging pups.

Six or seven days after dropping the pups, the females are ready again for mating. The males become agitated and fights begin as the younger males challenge the mature bulls for supremacy. The bulls in the prime territories mate with the cows, but often beachmasters are toppled from their thrones, and so each cow is likely to mate with several different bulls. Mating lasts for about 10 to 15 minutes and is brought to an abrupt end when the female bites at the male's throat and swings away her hindquarters.

In order to produce sufficient milk for the pups, the females must go to sea to feed. Each time they must re-run the gauntlet of solitary males. At sea, each mother spends two days catching squid. The milk she produces is five times richer than cows' milk and her youngster will double its birth-weight in its first month. The pups, meanwhile, gather into crèches and keep well clear of the sparring bulls.

The fighting is highly ritualized. Bulls charge and then stop, slew their hindquarters at each other and stare obliquely with the head averted. Hooker's sealions, it seems, have evolved more sophisticated territorial behaviour than other sealions, and this greatly reduces the level of violence. If posturing fails, the males will fight fiercely, although they still use a standard set of rules. They exchange blows about the head and push chests, like Sumo wrestlers. Each bull waits for the other to lunge for an unguarded flipper. The wounds can be deep but most of the injured recover, their bodies protected by thick layers of blubber.

At the end of mating, territorial order breaks down rapidly and the young bulls invade the beach in a last ditch attempt at setting up their own territories at the sites where the females are gathered. By this time, though, they are exhausted and the females move up and down the beach unmolested.

The pups are allowed to wander freely and the pod of pups at the top of the beach gradually enlarges. Now the pups encounter their first danger. Rabbits were liberated on the island during the last century and their burrows can be death traps for young sealions. The youngsters are able to move about on the land propelled by their front flippers, but they can only move forwards. If their heads become stuck in a burrow then they cannot back out. Each season, one in ten of the pups becomes wedged and dies.

The pods of growing pups serve as training squads for the adolescent males to rehearse their role as future beachmasters. They herd the pups and fight over the pods, just like a mature bull would fight over a group of cows. The female pups seem to solicit the attention of the adolescents, while the males sink their teeth into each other's hides. They learn the rituals and strengthen their backs, limbs and jaws for a life at sea.

Hooker's sealions were practically wiped out during the nineteenth-century seal and sealion slaughters. Today they breed on Enderby Island and on just three other sub-Antarctic islands to the south of New Zealand. They are protected, but still they are the rarest sealion in the world.

The Sea Daisy

It lives on submerged logs off New Zealand, has no stomach, stores food in its feet, and has ten sexual organs – it is the sea daisy, a sea creature discovered at the end of 1985 that is not only a newly discovered *species* but also represents an entire new *class* of animal.

Plants and animals are divided, for the convenience of scientists, into clearly defined groups. An individual belongs to a *species* in that it is identical, to all intents and purposes, to others of its kind. Several species, with many attributes in common, can be grouped into a *genus*, and in turn, the genus may have characteristics similar to others and so constitute a *family*. Families in common become an *order*, and several orders can be grouped into a *class*. Classes are grouped into *phyla*, which can be assigned to a *kingdom*.

Man, for example, can be classified as follows: Kingdom – Animalia, Sub-kingdom – Metazoa, Phylum – Chordata, Sub-phylum – Craniata, Class – Mammalia, Order – Primates, Family – Hominidae, Genus – Homo, Species – sapiens.

The sea daisy has been assigned to the phylum Echinodermata – the 'spiny-skinned' creatures that include starfish, sea urchins, sea cucumbers, brittle stars, and sea lilies. But it was thought that all the classes of echinoderms had been found, so that to discover yet another is quite an event in biology.

The sea daisy's nearest relatives seem to be the starfish, although the resemblance is not immediately apparent. It received its vernacular name from its similarity in shape to a daisy flower. It is a flat, wafer-thin disc, up to ten millimetres across, with a spiny upper surface and an underside that has a thin membrane much like plastic film stretched across an upturned saucer. Around the periphery of the lower surface is a ring of tube feet, quite unlike the arrangement in other echinoderms that have them in rows based on a five-star symmetry.

Vestiges of a pentasymmetry can be seen in the positions of the ten gonads which take up most of the space on the remainder of the underside. In the gonads embryos have been found indicating some measure of parental care.

The sea daisy lives on bacteria that accumulate in the holes of waterlogged wood resting on the bottom of the sea about 1,000 metres (3,300ft) down.

Animals similar to the sea daisy have been found as fossils in rocks hundreds of millions of years old.

The Night of the Blizzard

Australia's Great Barrier Reef stretches from Bramble Cay in the northern Torres Strait to Lady Elliot Island in southern Queensland, a distance of 2,000km (1,243 miles) (greater than the length of the British Isles), and covers an area some 200,000sq km (77,220sq miles), much of it below water. It is the largest reef of its kind, the next longest being off the coast of New Caledonia. More correctly it should be considered as a collection of reefs, islands, inlets and shallow bays, with limestone hills, plateaux, valleys and chasms, all built from an accumulation of the minute calcium skeletons of coral polyps and other marine organisms. Here and there, granite outcrops form more permanent steep-sided islands, but the land and seascape is dominated by the coral. Joseph Banks was on Cook's *Endeavour* when it struck the reef at 11 o'clock on the night of 10 June 1770. He wrote down his first impressions:

> It is a wall, rising perpendicularly out of the unfathomable ocean, always overflown at high water commonly seven or eight feet, and generally bare at low water. The large waves of the ocean, meeting with so sudden a resistance, make here a most terrible surf, breaking mountain high.

To the west of the reef is the Queensland coast. To the east is the Pacific Ocean that drops to a depth of 2,000m (6,562ft) not more than 7km (4 miles) from the edge of the reef.

The fauna and flora vary considerably throughout the reef. The northern section is in tropical latitudes while the southern tip is cooler, and so the further south you travel the less variety you see. On some of the islands wedge-tailed shearwaters or muttonbirds make their burrows, a hazard for any researcher creeping about during the night, and on others grassy beach platforms provide nesting sites for common noddies.

Heron Island, with its forest of pisonia trees, is host to white-capped noddies and reef herons, while Green Island, covered by mixed vine forest, has nesting Torres Strait pigeons and roosting flying foxes. Below the waves multicoloured coral reef fish, enormous groupers, gaudy starfish, and ugly sea cucumbers vie for your attention. In the deeper waters, migrating humpback whales compete with tiger sharks and manta rays for the impact of size and underwater perfection. Both in and out of the water, six of the world's seven species of marine turtle are to be seen in their thousands. On Raines Island alone, a count of nesting green turtles on just one evening revealed a total of about 10,000 individuals. Green turtles and loggerheads are so numerous that some of the largest aggregations in the world are to be found on the Great Barrier Reef.

The green turtle, with its horny serrated beak, feeds on algae and sea-grasses. The loggerhead's beak of thick plates is able to crush molluscs – ear shells, horn shells, turban shells and clams. Together with shellfish-munching stingrays, coral-crunching parrot fish, and cyclonic storms, the loggerheads contribute the smashed calcareous debris which has gradually accumulated to form the reef itself.

The animal responsible for the Great Barrier Reef is the coral polyp, a small sea anemone-like creature that sits in a partitioned calcareous cup. The cup is secreted by the coral and it is this that is left behind when the coral dies. A reef grows as more and more coral polyps grow on top of their predecessors. They feed by catching small zooplankton or other food particles with a ring of tentacles around the mouth. The reef-building corals, though, may gain extra nutrients from their live-in guests, tiny green algae, known as zooxanthellae, that live actually in the coral tissue. Here they make use of waste nitrogen from the coral and also make their own food by photosynthesis, using sunlight, carbon dioxide and water. The reef-building corals can live therefore only in shallow water, where sufficient light can penetrate.

Each polyp can be male, female, or both, and can divide asexually to produce more of its kind or can reproduce sexually, involving the production of sperms and eggs in testes and ovaries in the body cavity. It is the production of the eggs, sperms or tiny bundles of both that has caused a great deal of surprise and excitement amongst Australian marine biologists, for they have discovered that many of the corals, the length and breadth of the entire reef, spawn at the same hour, on the same night, in the same month. The process is synchronized with sea temperature and the phase of the moon. The whole lot 'go-off' at exactly the same time, and researchers such as Peter Harrison and Bette Willis of James Cook University, Townsville, can predict the event almost to the very minute. It happens in the week following full moon, either in November or December, when water temperature is increasing. If the water temperature rises rapidly then the event takes place after the November full moon, and if slowly then it's in December. So, a combination of factors is at work. There is also a light-dark component, for it has been found that some corals spawn at slightly different times after dusk.

Says Bette Willis:

If you were in an area with lots of staghorn coral on the night they were to spawn it would be quite spectacular. If you could look closely at the polyps you would see small pink bulges underneath the mouth, and these are the eggs gathering up prior to release. Most eggs are pink although some species have tan eggs and others green. You probably have about ten minutes to an hour of warning, then, all of a sudden, these egg masses, with sperm tied up amongst the eggs if it is a hermaphroditic species, shoot out into the water. The effect is like a snowstorm, except that all the pink bundles are floating up to the surface – it's like a blizzard with the snow going upwards. It's quite an experience. From a boat you look down on the surface of the water and see a soup of gametes.

Small crabs and fish are thought to take advantage of the mass release, and it may be that this event is a means of satiating predators to maximize the survival of the coral larvae. The larvae develop within 24 hours, and they swim freely in the ocean for about a fortnight before settling down to start another reef.

The Mermaids of Shark Bay

The high-pitched whine of a speeding powerboat announces its arrival on the flat calm of Shark Bay on the west coast of Australia. Ahead, in the water, there are dotted numerous bobbing brown heads, peering inquisitively at the creature causing all the disturbance. They do not move. The boat, its occupants seemingly unaware of the living obstacles ahead, ploughs on. They still do not move aside, seemingly unaware of the danger heading towards them. At the last moment the boat veers to one side. They are safe.

In fact the animals were safe all along. The speedboat was just part of an experiment being carried out by Canadian researchers, under the leadership of Paul Anderson of the University of Calgary, who had travelled across the world to study the 'mermaid of the sea' – the dugong or sea cow.

The dugong is a marine mammal related to the manatee. It breathes air as we do, suckles its young, but lives its entire life in the sea. Comparatively little is known about dugongs, for they are shy and retiring animals. They have, however, featured much in the folklore of the sea. The dugong is thought to be the animal that gave rise to the myth of the mermaid. Separating fact from fantasy in the literature is difficult but one disturbing piece of information has emerged. Dugongs, according to the Canadian scientists, are unable to detect objects, such as boats, that move faster than 25 knots. They are therefore unable to take avoiding action and are in grave danger of being hit and killed or mutilated in collisions with sports-fishing boats or the fast speedboats of water-skiers. The implications for their near relations, manatees, living in the busy waters around Florida's holiday resorts, are understandably serious.

On overcoming their shyness, the dugongs of Shark Bay were found to be very inquisitive creatures. Often they would surface, placing themselves between the sun and the scientists' boat, and carefully investigate the nature of the occupants. They might come within a yard of a man in the water, but having satisfied their curiosity, they would swim slowly away following a zig-zag course, glancing back first over one shoulder and then the other.

Dugongs and manatees are unusual as sea mammals go. They are herbivores. All the rest – the seals, dolphins, whales, polar bears and sea otters – are carnivores. The dugong feeds exclusively on sea grasses. These are flowering plants that grow in shallow sheltered bays, where light can penetrate and where wave action is minimal. The dugong strips the submarine foliage or digs up the rhizomes, curiously without even raising the sediment. Any grit or sand that does get into the food is carefully washed away before it is consumed.

Although closely related to the manatee, the dugong differs fundamentally in the shape of its tail. The manatee has an enormous rounded, paddle-shaped tail, whereas the dugong possesses crescent-shaped flukes, giving the animal the appearance of a cross between a large seal and a small whale.

The clue to the ancestry of both species is in their herbivorous diet. The abrasive plant food wears down the teeth and each worn tooth is replaced by another from behind. The only other mammal group in which this occurs is that containing the elephants and the hyraxes. About fifty to sixty million years ago dugongs, manatees and elephants had a common ancestor. The fossil record also reveals that dugongs and their relatives were once very common throughout the world – almost an 'age of dugongs'.

For the last century the dugong has been confined to the Indian and Pacific Oceans – along the east coast of Africa, in the Red Sea and Persian Gulf, around the coasts of India, Sri Lanka and New Guinea, and in isolated bays along the western and northern coasts of Australia. In all areas it is considered to be threatened by the activities of man. Only recently the dugong has been caught up in the war between Iran and Iraq. Huge quantities of oil were spilled from damaged oil installations into the Persian Gulf. In one survey, in 1983, 50 dugongs were found dead and these were thought to represent a large chunk of the total population living in that area.

Western man discovered the dugong in the late eighteenth century, but that encounter marked its near extermination. Until then, the dugong was known only to the peoples of the Near and Far East, and limited subsistence hunting by coastal aborigines was not detrimental to their numbers. Along the coast of northern Australia settlers and explorers witnessed herds over 5km (3 miles) long and 275m (900ft) wide, but that is rare sight today.

Unfortunately for the dugong, its meat is good to eat, tasting much like veal. The mid-nineteenth-century seal and whale hunters preferred dugong to whale and seal, and the animals were slaughtered in their thousands. Dugong meat, though, has been relished by many. In Sri Lanka, for example, it is consumed in certain religious ceremonies, and the Muslim community is allowed to eat it instead of pork. The thick skin can be tanned to make a tough leather, suitable for the bottoms of sandals. There is even the story that dugong hide was used to cover and protect the Ark of the Covenant on its journey across the Sinai Desert.

The teeth, ground into a powder, provide traditional 'cure-all' medicines, and in Madagascar it was said to be effective in curing food poisoning. Dugong teeth strung on a necklace are still thought by some people to have magical protective properties, although in some villages they are simply cut and polished and made into cigarette holders. Smoking a cigarette through a tooth is thought to invigorate the sex drive. The dried penis is much sought after also as an aphrodisiac. Fat from around the head was used to alleviate headaches, and oil, from the blubber, was supposed to be good as a laxative. Nothing, it seems, was left over; even the bones were used in making charcoal.

Catching and killing dugong was, and still is in some areas, an unsavoury business. An animal is harpooned from a small boat and after a brief but violent struggle is trussed to one side. One of the crew jumps into the water and plugs the nostrils with corks; the creature is left to drown or suffocate. Young dugongs are easily caught after their mothers have been killed. They are kept alive and their 'tears' collected to be sold as *air mata duyung* to bring good luck, prosperity and success with women to those men who carry a swab of tear-soaked cotton wool.

It is thought that the slaughter at the beginning of the century, despite the slow recovery of populations, may have substantially altered dugong behaviour. Before the mass killings, dugongs gathered in enormous social groups. There was safety in numbers, and they could be approached with considerable ease. The few that survived the harpoons and the clubs lacked this security of numbers and as a consequence they and their descendants have become extremely shy animals.

Little is known about their social and reproductive activities. Nuzzling, prodding and muzzle 'kissing' have been observed, which are thought to have some relevance to either courtship or group-bonding. Pregnant cows give birth to a single calf which stays close to its mother, usually swimming over her back. Sometimes it will actually clasp her and hold on for a free ride piggy-back-style. When she surfaces to breathe, the calf slips off to one side and waits by her flipper, returning to its station when she submerges again. It stays with its mother for about a year and then, according to some reports, it joins a group of adolescents. Maturity is not achieved until the youngster reaches its eighth birthday. All being well, it might live to sixty years.

Contrary to mariners' tales, a female dugong does not hold her baby in her flippers in order to let it suckle, although it is this supposed behaviour that gave rise to the mermaid myth. The legends may have gained even more credibility when dugongs were used as surrogate human females in initiation ceremonies, the likes of which are still thought to take place around the New Guinea coast.

Coastal peoples have an extensive mythology involving dugongs. Many tales relate to people who have violated taboos and have been thrown into the sea only to be turned into these mournful and pitiful creatures.

Mermaids, as such, probably had multiple origins. Norse mythology has its mermaids, but these are more probably directed at the dugong's close, but now thought to be extinct, relative known as Steller's sea cow. This enormous dugong-like animal lived in the more northerly waters of the Pacific. It was not discovered by science until 1741, but its flesh, like that of the dugong, was considered palatable, and so it was killed mercilessly until, in 1768, it was thought to have become extinct. There is some speculation that it might still survive in the ocean. In 1983, for example, a skeleton was found on a Soviet Pacific island, the first to have been found this century.

9
The Living World of Sharks

'Man is separated from sharks by an abyss of time', wrote underwater explorer Jacques Yves Cousteau in his classic *The Silent World*. But increasingly today man and shark are to meet, as scuba divers and snorklers, sports fishermen, surfers and wind-surfers – all the products of an expanding leisure market – invade the sharks' domain. The long-time human fear of these creatures stems from centuries of ignorance. 'Shark fever' accompanies most encounters, but it is, perhaps, sobering to find that more people die in road accidents during a bank holiday weekend in the USA than have been killed by sharks worldwide in the last ten years. Even when visiting an area frequented by sharks you are in far more danger on the way to the beach from bee stings, snake bites and lightning than you are of being the victim of a shark attack.

But the fact remains that sharks *do* attack people and these attacks are quick and terrible. Sharks, though, are generally not *man-eaters*, but *man-attackers*, a subtlety lost on their victims. Why sharks should behave in this way is only slowly being understood, for the life of sharks, until now, has been a complete mystery.

Where do sharks go, and how do they find their way? Do they prefer particular conditions of temperature or salinity? Do they communicate with other sharks of the same species? Do they lay eggs or give birth to fully formed young? Gradually marine biologists from all over the world have been piecing together a jigsaw of shark observations about courtship and mating, distribution and migration, locomotion, aggression, social behaviour, and the way sharks perceive their underwater world.

Contrary to popular belief, sharks are not 'primitive' killing machines, but are the sophisticated product of 350 million years of evolution. They have the most advanced capabilities. Sharks see well, and in colour. They learn intelligence tests as fast as rabbits. They can orientate and navigate using the earth's magnetic field. They can detect the minute electrical fields produced by the muscles of their prey. Some give birth to fully developed young. Some sharks are so streamlined that they create very little hydrodynamic noise, and swim so efficiently that they avoid acoustic detection by the prey they are chasing. Colour patterns of many species are such that the animals are functionally invisible. Most sharks are at the end of their food-chain. They have few enemies, save other sharks and man.

They have few parasites or diseases. To the biomedical researcher, sharks have become the focus of serious medical investigation. Early work on the immunology of sharks has revealed that they have antibodies that can fight most bacterial and viral diseases. Cancer is almost unknown. Current research is attempting to identify anti-carcinogenic factors in the blood of sharks and rays that may help in the fight to find a cure for cancer in humans.

Sharks are a successful group of animals that evolved from primitive armour-clad fish about 350 million years ago. Individuals very similar to those swimming today are to be found in rocks 160 million years old. They saw the dinosaurs come and go. About 300 million years before the present, there was a 'golden age' of sharks when this group of fishes was first to diversify and fill every ecological niche in the sea. There were those with claws like crabs, others with wings like flying fish, one species with a hinged dorsal fin like a ship's rudder, and another with spines on its head. Most of these fossil finds, from Bear Gulch, Montana, and Bearsden, a suburb of Glasgow, are little more than a metre long, and one less than an inch. Most died out, either ousted by the evolving bony fishes (the shark skeleton is made of cartilage), or pushed into extinction by some cataclysmic event.

One monster that survived until about 15 million years ago was a gigantic version of the great white shark. Its 15cm (6in) fossil teeth indicate an animal 14m (45ft) in length, probably the greatest predator to have lived on earth. Reconstructed jaws in the American Museum of Natural History have a gape in which six full-grown men can comfortably stand.

Today, about 350 species of sharks survive, including 'living fossils' such as the strange horned goblin shark *Mitsukurina*, an animal thought to have been extinct for 100 million years until it was found alive and well swimming off the Japanese coast.

Extant sharks form a relatively small group of fishes, and are found at all depths, in almost every ocean and sea in the world. Some, like the bull sharks, are to be found migrating regularly to and from freshwater lakes and rivers. They are found in Lake Nicaragua, Lake Jamoer in New Guinea, and Lake Izbal in Guatemala, and up and down the Amazon, the Ganges, the Zambesi and the Mississippi. It is not known how their bodies cope with extremes in salinity.

Off the Yucatan Peninsular and around the Japanese coast, reef and requiem sharks demonstrate another salinity puzzle. In certain under-water caves, known man-eaters can be approached and even caressed without fear of attack as they lie motionless, apparently comatose, on the cave floor – they give all the appearance of 'sleeping'. Fresh water, containing high levels of oxygen, is thought to seep into the caves and allow the sharks to breathe without having to pass water over their gills. The water also loosens parasites which are gobbled-up by 'cleaner' fish living in the caves – it is the shark equivalent of a drive-in carwash and valet service!

Sharks vary considerably in size, from the largest fish in the sea – the harmless, plankton-feeding whale shark, which can reach a length equivalent to two buses – to the smallest of sharks, such as the sneaky semi-parasitic cookie-cutter shark, no longer than a pencil.

Sharks also come in a variety of shapes, each adapted to a particular way of life. There are the sleek, torpedo-shaped ocean-swimming mackerel sharks, the great white, mako and porebeagle, the flattened bottom-feeding angel sharks, the highly manoeuvrable hammerheads, and the threshers with their long sickle-shaped tail for herding shoals of fish.

The greatest mystery, perhaps, in the life of sharks is where they go and how far and fast they travel. Do they swim in groups or are they solitary? Do they go on regular migrations? To answer these questions, scientists have enlisted the help of sports fishermen. They tag captured sharks and release them back into the sea. Already the researchers have found that blue sharks travel up to 5,794km (3,600 miles) from the release point, swimming about 20–30 miles a day. It appears that most blues follow the North Atlantic Gyre, a clockwise movement of ocean currents that takes them with the Gulf Stream across to Europe and then back to the Americas via the North Equatorial. But not all go on the journey. The eastern Atlantic fish turn out to be smaller than those in the western Atlantic, and there is a segregation by sex. Males stay close to the coast of North America, while the females migrate across the Atlantic to Cornwall, the Azores, and the Canary Islands.

Where many sharks go seems to be governed partly by the water temperature. Blue sharks prefer it cold, while tigers and hammerheads stay in warmer waters. Along the northeast coast of America each summer, the arrival of the different species of shark is determined by the position of the 20°C (68°F) isotherm. Blues stay ahead of the isotherm, makos come next, followed by the tigers and hammerheads. In the autumn they all head for deep waters again.

The movements of basking sharks – the second largest fish in the sea and, like the whale shark, a harmless filter feeder – has fascinated scientists and fishermen alike. They frequent the northeast Atlantic, coming inshore along the western coasts of the British Isles during the summer months. Then they just disappear. It was speculated that they migrate south to the Mediterranean, but recent studies suggest that they 'hibernate' on the bottom of the sea, at the base of the continental slope, and moult their gills. Attempts have been made to track them by satellite but with little success for it is very difficult to get a transmitter to stay attached to the shark.

And, how does a shark, swimming in what appears to be a featureless ocean, know where it is and find its way to where it wants to go? It seems that like many other widely travelled creatures, they use the earth's magnetic field.

Sharks are the ultimate hunters. At 400m (¼ mile) a great white can smell blood or body fluids in the water, and follow an olfactory corridor to head upcurrent towards a victim. It can compare minute differences in the current flow on either side of its body with sensory information picked up by the lateral line. It can also hear and accurately locate the low frequency sounds or vibrations of commotion in the water. At 23m (75ft), and almost in the dark, it can see the movement of prey, and in colour. The shark's eyes are ten times more sensitive to dim light than are a human's, and it can discriminate between blue, blue-green and yellow. If heading rapidly to the surface, perhaps when attacking a seal from below, the shark can switch off its dark-adapted system and thereby function normally in bright light.

When it closes in for the final attack, a membrane covers and protects the eye from struggling prey, and the shark is essentially swimming blind. At this point a remarkable sensory system comes into play – one that detects electricity. Sensory organs, located in small, jelly-filled pits in the snout, can detect minute electrical currents associated with muscle activity, such as a beating heart or the movement of fins, in the prey. In tests, sharks have been able to find flatfish buried in the sand by picking up the electrical field around the fish. You may have noticed in natural history films about sharks on the television that great whites often try to take a bite out of the scuba divers' protective cage. It may look as if the shark is attempting to get at the divers inside but in reality it is just responding to the electrical field generated by the metal of the cage in the water. In response to the stimulus it does the only thing it is programmed to do, and that is to open its mouth and attack.

The shark's bite is powerful, but the main damage is caused by the shearing action of the teeth. It has an endless conveyor belt of teeth, at various stages of development, which are probably totally replaced about five times a year. Great white sharks have even, triangular-shaped, razor-sharp teeth in both the upper and lower jaw. They are used for tearing off chunks of blubber from dead whales or seizing and chomping marine mammals such as seals and sealions. Its close relative, the mako shark, on the other hand, has longer grasping teeth for grabbing fish.

Miniature teeth, all over the shark's body, giving it a skin texture like sand paper, trap mucous. This serves to present less resistance in the water. Swimming is further enhanced in the great white by keeping the powerful swimming muscles a few degrees warmer than the surrounding seawater. At the Woods Hole Oceanographic Institution researchers have been using radio tags to follow great whites and have found that the muscles are kept at 7–10°C above the ambient seawater temperature. This is important for the shark as it has been shown that for every 10°C rise in muscle temperature there is a three-fold increase in muscle power. The shark achieves this with a special blood supply system to the swimming muscles. In dissection it looks a little like an old-fashioned central heating radiator and acts as a heat exchanger. Warm blood is prevented from being carried off to places such as the gills where heat would be lost to the outside. It is significant that the three species of shark showing this heat-retention system are some of the most powerful fish in the sea – the great white, the mako and the porebeagle.

Courtship and mating is not a gentle affair. Females are often seen with badly slashed bodies and torn fins. Mating itself, in any species of shark, has only rarely been seen, although the process in sharks is more sophisticated than in the bony fishes. All sharks copulate and fertilization is internal. Embryo development can be any of three ways. Some sharks lay tough, horned eggs known as 'mermaids' purses'. Others retain the embryos in the female's body, and give birth eventually to well-formed youngsters. Still others develop a placenta between the embryo and the mother's uterine wall. For some species, notably the sand tiger, competition *inside* the female is fierce. Some babies eat their womb-mates, and then go on to gobble the eggs that are continuing to be produced by the female.

Sharks are thought to be more intelligent than we ever imagined. They have been taught to ring bells, negotiate mazes, and discriminate a variety of stimuli. Recent research on the shark's brain has shown it to be larger and more complex than was previously thought. In some tests, sharks have been shown to solve problems in the same time it takes laboratory rats.

Observations of over 2,000 shark attack incidents worldwide showed that sharks have two types of attack. One is when it is simply feeding, but the other, and the more common, is reminiscent of a threat. About 75 per cent of shark attacks appear to be the result of sharks *threatening* the human invader. Could it be that sharks perceive man in the water as some kind of competition? The grey reef shark, when approached too closely, gives an identifiable threat display – it curves its body into an

S-shape, and just hangs in the water – much like a cat arching its back. If a human diver continues to advance, it eventually stops displaying and races in with an attack. The mouth is opened and the shark slashes at its adversary in an attempt to frighten him away. The discovery that a shark has a threat display must mean that it reacts socially to other sharks. It is likely that a whole new 'shark language' is waiting to be unravelled. It seems they 'talk' to each other more than we first thought, albeit with a system of postures.

The great white shark, otherwise known as the 'man-eater', is not as reliable in indicating its intent to attack as the grey reef. The only predictable thing about a great white is its unpredictability. They attack anything, at any time, and without warning. It is the league leader in shark attacks on humans, and each attack is front page news.

A shark attack seems to trigger off that primeval fear we have of creatures from the deep. Newspapers worldwide will print the story and for some reason we are compelled to read and absorb every terrifying detail. It is like watching a horror movie. You are scared but remain glued to the screen, hiding the eyes only in the really gruesome bits. Psychiatrists suggest that much of the concern with dangerous animals stems from Man's shadowy and distant past when our ancestors came down from the trees and faced a most terrifying threat – that of being eaten alive. The shark, more than any other mankiller, bring out this ultimate of fears. Who can argue when you see those cold expressionless eyes, a mouth bristling with rows of serrated teeth, and learn that the creature might grow to 12m (39ft) long and swallow a man whole?

In reality, of the 350 recognized species of shark, only 30 have been known to deliberately attack humans. Of those, the great white, the tiger, the hammerhead, the bull, the grey nurse, the oceanic white tip, the mako, and the blacktip figure strongly in the records, and most find human-flesh so unpalatable that they spit us out.

David Baldridge, of the Mote Marine Laboratory in Sarasota, Florida, has studied the most complete collection of shark attack statistics in the world – the International Shark Attack File. It was kept between 1958 and 1969 by the US Navy and the Smithsonian Institution in Washington DC, and it contains nearly 2,000 entries of instances when sharks have attacked humans. In his analysis, Baldridge found that the frequency of shark attacks is relatively low: 'Recorded shark attacks are no more than 20–50 a year', he stated, 'and out of those only 20% are fatal, that's only 10 fatalities a year worldwide as a maximum.'

'Shark attacks can occur in any depth of water,' concludes Baldridge, 'but, in fact, statistically most shark attacks have been in water that is waist deep or less. You might think that that is the most dangerous area of water to be in, but in truth, that happens to be where most of the people are standing, swimming or playing.'

The real risk of shark attack increases as you move further from the shore: the deeper the water, the more you are in the shark's domain. Curiously, this seems to be linked to the greater number of attacks on men than on women. The ratio of male to female victims off beaches is about 8:1. Baldridge is inclined to believe that the higher number of attacks on males is tied in with the way the human male behaves: 'The male on the beach is more likely to be involved in aggressive, active behaviour, such as splashing and playing around, but he also swims furthest and is likely to be one or a group of males on the outer fringes of swimmers, and so is more accessible to a passing shark.' Baldridge's hypothesis gains more credibility when the records are examined and are limited to incidents outside a line 120m (130yds) from the shore: the male to female ratio goes up to 30:1.

Although you are at more risk the further out you go, the records show that over half the attacks on beaches occur within 60m (66yds) of the shore, the zone where most of the bathers are to be found. Ed Broedel, a colleague of Baldridge, counted people at Myrtle beach, South Carolina, and found that the distribution of bathers strongly tied in with known shark attacks. For example, 17 per cent of bathers were seen to be wading in knee-deep water, and 16 per cent of victims have been in water less than knee-deep. Broedel found that 73 per cent of all bathers swim in water up to neck deep, and Baldridge discovered that 78 per cent of victims have been struck in those waters. The depths at which sharks attack, suggests Baldridge, are related to nothing other than the human population distribution and to no significant behavioural or physiological reason on the part of the shark.

Temperature is another factor that is often considered. It has been suggested that the water must be over 21°C (70°F) before a shark will strike a human. Baldridge, together with Ed Broedel and Beth Arthur, looked at their beach populations to see if there was a simple explanation. Sure enough, they found that people are more reluctant to go swimming in water below 21°C, so that there would be fewer people in the water for sharks to attack. Once again, it is human behaviour and not shark behaviour that is significant in the numbers of shark attacks.

Chemicals to repel shark have, so far, proved to be ineffective; that is, until a flatfish – the Moses sole fish – was found to be particularly repugnant to sharks. A shark about to take a chunk out of a Moses sole will stop in mid-bite and drop to the bottom of the tank, paralyzed.

Sharks have more to fear from us than we have to fear from sharks. We eat a lot of sharks. Even in the British fish and chip shop shark meat is frequently eaten as 'rock salmon', a euphemism for the common dogfish. In the Far East the tope donates its fins for soup, and in Mexico salted shark meat is eaten by the poor. 'Squalene', derived from the livers of deep-sea sharks, is used as a base for lipstick and a cure-all in Japan. In the search for a remedy against constipation or a cure for cancer, some Japanese will pay the equivalent of a pound sterling for a small capsule. It is known as 'marine gold'. The jaws of a large great white will sell for over $1,000, leading some scientists to declare it a threatened species.

Surprisingly, new species of shark, even very large ones, are still being discovered. Storytellers would like us to believe there are enormous, unknown fish swimming in the inaccessible parts of the oceans, and the truth of the matter is that there are.

On 15 November 1976, the US research vessel AFB-14 was about to get under way from its station about 42km (26 miles) northwest of Kakuku Point, on the Hawaiian island of Oahu, and was hauling in two enormous parachute-like sea anchors from a depth of 165m (540ft), when the crew realized that they had accidentally caught some large creature of the deep. Trapped inside one of the 'chutes was a very large shark, about 4.46m (14ft 7in.) long and weighing 750kg (1,650lb). Its great blubbery lips surrounding a broad gape set on protruding jaws instantly gained it the nickname 'megamouth'.

The strange fish was hauled on board, with considerable difficulty, and taken to the Naval Undersea Center at Kaneohe Bay. The following day it was examined by Leighton Taylor, of the Waikiki Aquarium who realized that it was something quite new to science. On hauling it out on to the dockside, the tail broke and the carcass fell back into the water, but Navy divers were able to locate and recover it. Hawaiian tuna packers in Honolulu put it into deep freeze until the scientists could work out what to do about preserving it for examination. Two weeks after it was caught it was thawed and injected with formalin. Leonard Compagno of San Francisco State University and Paul Strusacker joined Taylor to describe the animal officially.

They gave it the scientific name *Megachasma pelagios*, meaning large-yawning mouth of the open sea. It is thought to be a filter feeder, although morphologically and anatomically different to the other two filter-feeding sharks of the open ocean – the basking and whale sharks – and has been grouped uneasily with the lamnoid sharks in its own family, the Megachasmidae.

The skeleton is composed of soft cartilage, and the muscles are thought to be sufficient for slow, steady swimming. Unlike the basking and whale sharks which skim plankton from the surface waters, megamouth probably swims, with jaws agape, through patches of euphausiid shrimps that live in the semi-darkness 150–500 metres

(490–1,640ft)deep. There is speculation that shrimps are enticed towards the mouth with the aid of bioluminescent spots around the mouth. Silvery tissue, dotted with small circular pits, was found lining the blubbery mouth. Inside is an enormous tongue and closely-packed finger-like gill rakers. It may be that the tongue is used to compress the water and shrimps against the roof of the mouth and through the gill rakers in order to sieve out the food, which is then swallowed. There are also 236 tiny teeth that, more than likely, help with the filtering process. In the stomach the researchers found a soup of shrimps, and further down the gut a new species of tapeworm.

Externally, the mouth apart, the body is rather similar to that of a slow-swimming shark. The tissues are flabby and the tail fin is asymmetrical, indicating a creature not in a hurry, and typical of fish species that live in the nutritionally poor open ocean. On the skin were small circular weals, evidence of attack by the semi-parasitic and voracious cookie-cutter shark – a 50cm (20in.) long shark with the largest and sharpest teeth for its body size of any of the sharks. This small demon takes small circular bites from any slow-moving marine creature, including whales, sharks and large fish, by approaching its victim head-on, attaching to the body with its sucker-like mouth, and then, aided by the slow forward motion, swivelling around while gouging out a circular chunk of meat or blubber.

In November 1984, exactly eight years after the Hawaiian fish was caught, another megamouth was netted by the commercial fishing vessel *Helga* in a gill net set at 38m (125ft) off Catalina Island, near Los Angeles. Fortuitously, a California Department of Fish and Game observer was on board and he immediately recognized the fish as something special. It was transferred to the Department's own research vessel and shipped direct to San Pedro where scientists from the Los Angeles County Museum were waiting to examine it. About 700kg of ice was taken to the museum car park where the shark resided until a temporary case could be built. It was then moved into a fibreglass display tank containing 2,273 litres (500 gallons) of 70 per cent ethanol.

In an interview with the *Waikiki Beach Press*, Leighton Taylor was reported as saying: 'the discovery of megamouth does one thing. It reaffirms science's suspicion that there are still all kinds of things . . . very large things . . . living in our oceans that we still don't know about. And that's very exciting.'

Interestingly, both specimens of megamouth are males. Female sharks generally grow larger than males, so might there be even larger megamouths living in the depths of the Pacific?

10
The Americas:

i.   North America

North America, excluding the Arctic regions, has several major geographical features with their own distinctive range of plants and animals. There are the two major mountain chains – the young, sharp-peaked Rockies in the west and the older Appalachians in the east. Between them lie the Great Plains with three extensive river systems – the Mississippi–Missouri running to the south, the Mackenzie flowing north, and the St Lawrence and the Great Lakes going to the northeast. There are the alligator swamps of the Everglades in Florida, the arid deserts, including Death Valley, between the Rockies and the Sierra Nevada, and the mild West Coast tempered by weather systems originating in the Pacific Ocean.

Many of the animals of North America arrived as a result of migrations across two former land-bridges – the isthmus of Panama to the south and the Bering land-bridge in the north-west. As in Europe, the North American continent sees a constant seasonal movement of birds from south to north in the spring and vice versa in the autumn. Some believe that the birds follow distinct migration corridors, with recognized refuelling stops on the way.

Grizzly

In Wyoming's Yellowstone National Park, during 1983 and 1984, the grizzly bears were behaving in a mysterious way. For the most part they avoid contact with humans, only coming close to camps or garbage dumps in order to take scraps of food. If certain basic precautions are taken, such as cooking 180m (600ft) away from the tent site, taking off the clothes in which you cooked before returning to the tent, washing-up all cooking utensils, and locking all left-over food and rubbish scraps inside the boot of the car, then the bears should leave you alone.

On 5 August 1984 a Californian family complied with all the park recommendations and regulations, but as dusk came, a 3m (10ft) long bear rampaged through the camp site and seized a twelve-year-old boy. Fortunately he struggled free, but he was badly mauled. A few days later a Swiss hiker was killed in her tent by a bite to the neck. The following week a middle-aged couple were attacked but, despite deep wounding, were able to crawl to their car. The previous year a young camper in the Gallatin National Forest, to the north of Yellowstone, was dragged from his tent, killed and partially eaten. Why were the bears behaving in such a ferocious and quite uncharacteristic manner?

The answer, and it is only a suggestion at present, is that they were 'high' on drugs. Bears that make a habit of visiting rubbish dumps and frequent the camping sites are shot with a hypodermic and tranquillized before being moved to a section of the Park where humans rarely go. Several die-hards, though, have returned time and time again for the easy meals at the camps. So, time and time again they have been tranquillized and removed. Bear number 15 was thought to be the killer of the 1983 victim and it was shot and examined for brain abnormalities. What the scientists found, though, was not brain damage but residues of the drug PCP, known to pharmacists as phencyclidine and to the hippies of the West Coast as 'angel dust'. It is known to induce abnormal behaviour in humans, including homicide. It is also the main component of one of the tranquillizers used by the Yellowstone Park wardens.

The bears acquired their 'junk-food' habit many years ago when it was Park policy to feed the bears at spectator feeding stations. Visitors to the park were able to sit in a grandstand, surrounded by a cage guarded by armed wardens, and watch the grizzlies at feeding time. Lorries carrying vast quantities of waste food would drop their loads on to tables of wooden logs, and for forty years the bears came to depend on the sites for their daily meal. In the late 1960s, the policy changed and the bears were encouraged to find their own natural foods. Some, though, were hooked on the free hand-out. The worst offenders were shot. Between 1968 and 1973 about 200 bears were killed.

In the past, over 100,000 grizzly bears lived in the northwest states of the USA but after 180 years of persecution by hunters and farmers, fewer than 1,000 now survive. There are about 200 remaining in Yellowstone and the populaton is falling by an estimated 4 per cent each year. The grizzly bear is officially designated by the US Department of the Interior as 'threatened', and not as 'endangered', and so in the State of Montana up to 25 grizzlies a year can be shot.

Most bears are killed by farmers protecting their sheep, park wardens culling 'nuisance' bears, hunters who can't tell the unprotected black bear from a brown, and poachers. The pelt, teeth and claws from one grizzly can earn a poacher over $10,000. The teeth and claws are made into a perverse form of jewellery. Even the gall bladder is profitable. In southeast Asia the organ is proferred as an aphrodisiac.

The grizzly is also known as the brown bear, and was once widely distributed throughout the northern hemisphere. Its European cousin is the Eurasian brown bear, a subspecies found from northern Scandinavia to the eastern part of the Soviet Union, from Syria to the Himalayas, in the Pyrenees, Alps, Abruzzi and Carpathian mountains, and other remote places in Europe. In January 1983, five hunters disturbed a female with cubs and were attacked near the northern Greek village of Flanbouro.

On the north American continent, the largest brown bears are those found on the Kodiak, Afognak, and Shuyak islands in the Gulf of Alaska, where they are known as the Kodiak bear. Its Eurasian counterpart is the Kamchatkan brown bear which is found on the other side of the Bering Sea. The largest Kodiak in the wild, according to *The Guinness Book of Animal Facts and Feats*, weighed 750kg (1,656lb) and measured 4.12m (13ft 6in.).

Size varies with the time of the year and the availability of food. When it is plentiful an individual might consume as much as 16kg (35lb) in one day's foraging. The grizzly is an opportunist omnivore, eating mainly succulent vegetation, tubers, berries, insect grubs, carrion, the occasional rodent or deer, and salmon and trout when in season. The females have distinct home ranges, and although these may overlap to a certain extent, the resident animal will establish exclusive rights to forage there. Adult males are solitary.

Breeding occurs during May and June when male bears roam about in search of receptive females. Implantation of the fertilized egg is delayed until the autumn, when the female seeks out a denning site for the winter. This can be a cave or a hollow tree or a simple shelter beneath a fallen tree trunk. The cubs, usually two or three, are born naked and helpless between January and March. They may stay with the mother for up to four years.

When the youngsters grow up, the females may be allowed to stay in the mother's home range, but the males are encouraged to leave. Young males spend much of their early life avoiding other large males. Bears are naturally aggressive to other bears. As a female is only able to have cubs every three or four years, and therefore raises as few as six to eight during her lifetime, a premium is placed on every offspring. She will attack any intruders. With so few receptive females around, they too are at a premium, and males will chase out any immature pretenders muscling in on their patch. One problem here is that bears are short-sighted and a human wandering into a bear's territory may be taken for a competitor and attacked. It is possible that bear assaults on humans are simply cases of mistaken identity.

Bear fights are brutal and bloody affairs: often an animal is killed or badly wounded, and so they tend to avoid each other. Dominant males leave scent marks around their home range by rubbing themselves against tree trunks or making scrapes in the ground. At places where food is concentrated, such as garbage dumps or salmon runs, a large number of animals must necessarily come into close contact, and so a dominance hierarchy is established to ensure that fighting is kept to a minimum.

Grizzlies are large and powerful animals. They can stand on their hind feet and reach up to 3m (10ft) above the ground. They are surprisingly fast. Jack Olsen, author of *The Night of the Grizzlies*, estimated that a brown bear could outrun a human by 35 yards in a 100 yards sprint race. They can charge with a speed of 50kph (30mph). The forepaw can smash the head of an elk with a single blow. They are truly formidable animals.

Hibernating Black Bears

At the end of October each year, the black bears of North America retreat from the world and hibernate for the winter. They seek out a suitable den and for three months or more just go to sleep. Unlike other hibernating animals, the bears do not drop their body temperature down to the ambient temperature, but rather maintain it within a degree or two below the normal active body temperature. They do not eat or drink for the entire period and do not urinate. Females give birth and suckle their young during hibernation without having to leave the den.

The black bear keeps itself ticking over by breaking down stored fat and body protein. A byproduct of this process is the release of water which serves to replace that lost by evaporation from the skin and the lungs. In other animals a large-scale loss of water would be tied up with the excretion of nitrogenous waste, a product of normal metabolism, in the form of urine; but the black bear avoids this – it does not produce urine.

A build-up of urea in the blood should ultimately result in the animal's death – it is the same as kidney failure – and so the bear has got around this with a unique biochemical trick. The waste nitrogen is redirected to a new chemical pathway where it is rebuilt into amino acids that are incorporated into new proteins that the bear can re-absorb and re-use.

A research team led by Ralph Nelson, of the University of Illinois, also discovered that the transition from the active to the hibernation state is gradual. Bears were seen to stop eating and drinking some time before 'denning' took place, and when they came out again in the spring they did not indulge in a sudden orgy of feeding but instead sloped around for several weeks without showing any interest in food or water.

Ralph Nelson, a clinical nutritionist and director of research at Carle

Foundation Hospital, believes that the bears' method of dealing with waste products during hibernation might give us some clues about fighting human medical conditions such as kidney failure and obesity.

Mighty Mouse

The southern grasshopper mouse *Onychomys torridus* is an inventive and fearless nocturnal North American predator. It has been seen to outwit beetles that spray noxious chemicals, overcome stinging scorpions, and take lizards, birds and mice far bigger than itself. For each prey species it has a particular attack strategy.

It has even been seen to tackle a large and potentially dangerous grasshopper *Brachystola magna*. The aggressive mouse usually restrains scarab beetles, moths or small crickets and grasshoppers by closing in on their front end and simply leaping on them, despatching the victims with a bite to the head. If the mouse tried the same approach on *B. magna*, it would get a faceful of spines as the grasshopper brought its spiky and powerful hind legs up and over its head. And so the mouse avoids the grasshopper's defence by nipping around to the side and sinking its teeth into the femoral-tibial joint of the hind leg. With the leg immobilized, the mouse can go for the head and administer the *coup de grace*.

This is the latest in a lengthening list of predatory adaptations observed in the grasshopper mouse. *Onychomys* can similarly sneak up on a scorpion and immobilize its tail, and can quickly kill a lizard by biting into the spinal cord or crushing the cranium. It can deal with the chemically defended tenebrionid beetle *Eleodes* by grabbing it, holding it upright in its forefeet, and thrusting the quinone-spraying rear-end into the sand where it can discharge safely while the mouse devours it head-first. Whether these attack techniques are learned or innate is not clear, but they surely help the mouse to be an effective predator capable of taking a wide variety of prey.

Jaws of the Desert

Little is known about the adult ant lions – they are relatives of the lacewings and alder flies, are known as 'doodle-bugs' (long before the V-1 rockets of the Second World War), and fly at night only to live long enough to mate before they die – but their larvae are familiar to little boys from tropical and sub-tropical regions around the world. They are found in sandy areas, whether it be forest, dunes or desert, and they live below ground in pits.

The ant lion lives at the bottom of the pit, with just its enormous pincer-like jaws exposed to catch ants or other small insects that are unfortunate enough to fall in. The victim is grasped, pierced and sucked dry. The pit, on first inspection, appears to be a simple conical-shaped depression in the sand, but entomologist Jeffrey Lucas, of the College of William and Mary, Virginia, found that there was more to the construction of the pitfall trap than at first meets the eye.

The animal walks backwards in a circle, gradually spiralling inwards, loads sand on to its head, and then throws it to one side or the other, until the cone is complete. The largest pits can be up to 10cm (4in.) in diameter and 50cm (20in.) deep. The walls are steep so that it is very difficult for a trapped ant to climb out. In addition, the walls are lined with fine sand. Fine sand, it turns out, is far less stable than coarse sand and to take advantage of this property the ant lion sorts out fine sand from coarse as it excavates the pit. It does this by vibrating the forelegs at the sides of the head so that the larger particles rise up to the top and are tossed to one side and the smaller ones drop down to be retained in the pit.

There is also an asymmetry in the shape of the pit slope. The front wall has a steep slope, while the rear wall is shallower and is lined with the fine sand. If an ant attempts to escape up the rear wall the ant lion throws sand at it, one of the few examples of an insect using a tool. The wall below the ant, where the ant lion is removing sand, collapses and so the victim falls back into the pit.

They Come from the Sea

In late spring, the fishermen of Delaware Bay, on the Atlantic coast, begin to catch increasing numbers of curious crab-like creatures that look like animals from a by-gone age. They are the horseshoe or king crabs – not true crabs at all, but closer relatives of the scorpions and the spiders. They get their name from their shape: the hinged, protective carapace resembles a horseshoe. At the rear they have a long tail-like spine.

They spend most of the year foraging in the muds and sands at the bottom of the bay, but in spring they all head for the shore to breed. On the spring tides of early May, the males arrive first. They form a line several crabs deep and maybe a couple of crabs high along the edge of the sea. They are packed so tightly that tens of thousands of individuals can be found in a kilometre stretch. Along some of the undisturbed beaches the line of crabs can be uninterrupted for several kilometres.

As the tide turns, the females arrive. They seek places in which to lay their eggs, but on their emergence from the sea they become the focus of male attention. The 'thick black line' breaks up and the males clump around the females, each using its pair of special grasping pedipalps to 'lock-on' to a female carapace. Once attached, it is unlikely to be pushed off. Thus engaged, the female and male walk up the beach in search of a suitable egg-laying site. The female digs a hole, maybe 20cm (8in.) deep to avoid the probing bills of bird predators, and lays up to 80,000 eggs in regular bundles, some 3cm (1 in.) across and others 10cm (4in.). The male fertilizes the eggs as they are laid in the hole. In amongst the eggs the female mixes sand and pebbles, perhaps a further attempt at disguising the egg site. There are often so many animals laying eggs along the upper part of the intertidal zone that there might be as many as fifty nests on every square metre of beach.

Egg-laying complete, the male and female return to the sea on the next tide and disappear for another year. Their departure, though, marks the start of another event – the arrival of the birds.

Each spring, Delaware Bay becomes a 'refuelling stop' for millions of northerly migrating shorebirds, and it is the eggs of the horseshoe crabs that provide the food. From Brazil, Patagonia, Chile, Peru, Suriname and Venezuela they come, on their way to their breeding grounds in the Canadian Arctic. They are in a hurry. The first to reach the north get the best breeding sites. They arrive suddenly in the Bay, eat their fill, and depart.

Most birds probe about in the sand, plucking up single eggs that have not been buried deeply. The turnstones, however, go digging. All along the tide line they excavate the crabs' eggs, and squabble for possession. Birds with brightly coloured crowns win the day, and the food. If a hole is left for more than a minute the other birds jostle for position. Sanderlings often follow the turnstones and they, in their turn, fight to defend the prime feeding sites. Semipalmated sandpipers do not fight and several individuals may share a vacated hole. Laughing gulls patrol the beach and spot turnstones that linger too long at particularly rich sites, and then swoop in and take over. Even though each egg provides little nutrient in itself, there is such a glut of food that migrants can double their weight before they resume their flight north. It has been calculated that fifty thousand sanderlings consume six billion eggs weighing 27 tonnes in just a fortnight.

The timing of the birds' arrival and the emergence of the crabs is not coincidence. The migrants need 'staging posts' to facilitate their journey north. Delaware Bay with its horseshoe crabs is an important site. About 80 per cent of North America's *rufa* red knot (*Calibris canutas rufus*) population, for example, must use the Bay and depend on the crabs for refuelling. Land reclamation and beach stabilization programmes may alter the structure of the beach, destroying the nest sites. Spillage from tankers transferring oil to barges and pipelines, and leaks from the concentration of petrochemical works in the area, are a constant threat. One ill-timed spill could wipe out the crabs, the eggs, and the birds.

Horseshoe crabs have been on the planet for a long time. Fossils similar in form and shape to present-day individuals have been found in rocks estimated to be about 300 million years old. They are also remarkably similar to the trilobites, a large and diverse group of arthropods, that dominated the bottom of the sea from the Cambrian period to the Silurian (590–408 million years ago), and disappeared in the Permian. Indeed, the one-centimetre-long horseshoe crab larva, with its body divided into three sections, is known as the 'trilobite' larva.

The larvae reach maturity in their third year. The adult horseshoe crab's body is divided into two sections – the prosoma (the front bit) and the opisthosoma. There are six pairs of appendages on the prosoma – a pair of chelicerae or pincer-like mouthparts in front of the mouth and five pairs of walking legs behind. Similarly, there are six on the opisthosoma – a pair modified as a gill cover and five pairs flattened into leaf-like gill books for respiration. The movement of the appendages maintains a flow of water over their surfaces, and small crabs have been seen to swim along, upside down, using their gills.

The crabs live on sand or mud at the bottom of the sea. By using the tail spine as a lever to force down the front part of the body they skim the seabed. The fifth pair of walking legs are modified as shovels. Debris is caught by the pincer-like ends of the other walking legs and any food is passed forward to the mouth. They have also been known to crack open shellfish with the strong bases of the sixth pair of walking legs. Worms and clams form the best part of their diet.

The horseshoe crabs are considered to be 'living fossils'. The group first appeared in the Devonian period between 400 and 350 million years ago. Today, they are unique and have been assigned by taxonomists to their own order, the Xiphosura, and there are only four surviving species. They are to be found along the northeast coast of the United States and around the coasts of southeast Asia.

Back from the Brink

A walk through the sand dunes of the Ano Nuevo State Reserve can be an eerie if not slightly frightening experience. All around you can hear the low, guttural blasts, like reluctant diesel engines, of creatures that you have been warned can be two or three tonnes in weight, raise themselves taller than a man can stand, and move faster over loose sand than a man can run.

I had joined a small group of Californians who had driven the 96km (60 miles) south from San Francisco to see one of the earth's remarkable wildlife places and some of the animals that came back from the brink of extinction. Having paid our parking toll and ventured on to the dunes, we gingerly edged forward through a matting of wild strawberries, grasses and sea figs. A shadow passed overhead and we looked up in time to see a white-tailed kite hover, then dip, hesitate and swoop. A young brush rabbit scampered to safety and the kite rose again on a gust of wind from the sea.

We chanced upon a depression in the sand.

'One has been sleeping here,' said our student guide from the University of California at Santa Cruz.

There was a gasp from the onlookers. The hollow was nearly 4.6m (15ft) long. We crept apprehensively on to the beach and there it was – three tonnes of light brown and grey blubber – a fully mature male northern elephant seal, the largest of all the seals. He was about 5m (16ft) long, from the tip of his trunk-like proboscis to the ends of his hind flippers, and he was the boss around here – the beachmaster. Behind him were gathered his harem of slightly smaller cows and their pups. Languidly they scooped sand over their enormous bodies, perhaps as an aid to keep cool or maybe to reduce the irritation of skin drying in the constant onshore breeze.

Our party approached cautiously. We had been told that although the bull looks ungainly and helpless, when provoked he could be irritable and unpredictable, especially in the breeding season. His head could extend forward for a preliminary nip and his body could crush a frail human frame with one gigantic ripple of his gargantuan bulk. US Law, CAC 4325 – Closure Order No:2–3 – allowed us no closer than 6m (20ft), both for our safety and his welfare.

Unware of his statutory protection, the animal reared up, bent back his head and neck, inflated his nose to its full 23cm (9in.) in length, and blew half-a-dozen deep, throbbing snorts. He defied us to move closer. We all, as one, stepped backwards. I looked nervously over my shoulder: while we had stood mesmerized by this encounter I imagined another two-tonne monster shuffling up from behind, cutting off our escape.

Dotted amongst the dunes were many young bulls. They had arrived too early, confronted their contemporaries, fought hard and long, but were spent by the time the mature bulls and cows had arrived at the breeding beaches. We dodged between their exhausted, almost lifeless, bodies and clambered to the top of a giant dune where we could see the rest of the reserve.

Far out to sea we could see the spouts of migrating grey whales. In the next bay several Californian sealions were swimming not far offshore. In the shallow channel separating the mainland from Ano Nuevo Island, another large bull elephant seal wallowed in the shallow water, his hind flippers waving in the air like an admiral's pennant. The island, a small strip of Monterey shale isolated from the mainland by pounding surf and Pacific storms, was as recently as 150 years ago still part of the Ano Nuevo promontory. It was the island that was the first flipperhold for the returning elephant seals in 1955, some sixty years after their disappearance from the Californian coast at the time of the great seal carnage.

The story of the northern elephant seals of California is a remarkable one, for they were reduced in numbers so drastically that it is a wonder how they survived at all. They were slaughtered mercilessly during the nineteenth century for the oil that could be extracted from their blubber. The largest elephant seal known was 6m (20ft) long and rendered 955 litres (210 gallons) of high-grade oil from the fat stripped from its body.

For twenty million years, thousands of elephant seals had hauled out on to the islands and mainland beaches of the West Coast, but by 1892 just fifty to a hundred animals survived. They lived on the Isla de Guadalupe, off the coast of Baja California Norte, on a beach backed fortuitously by 914m (3,000ft) high sheer cliffs which allowed little access to the seal butchers. Most of this small population were safe, but not all. A dozen were killed and shipped to museums to be stuffed and displayed as one of the world's rarest species.

In 1922, the Mexican Government introduced legislation to protect these marine mammals. There was even an armed garrison to guard Guadalupe beach, now known as Barracks Beach. The US Government followed suit in 1930 and, gradually, individuals were found to be migrating from the enlarging Guadalupe population and returning to the other ancestral breeding grounds on Los Coronados, San Clemente, San Nicholas, Santa Barbara, San Miguel, Sant Martin, San Benito, Cedros, Natividad, Ano Nuevo, Farallon, and Point Rayes to the north of San Francisco. During the past fifty years the recovery of the populations of northern elephant seals has been nothing short of miraculous. In 1911 there were only six known births along the entire Pacific coast of North America. In 1982 there were an estimated twenty-five thousand, of which fifteen hundred were born at Ano Nuevo. This, according to scientists at Hubbs-Sea World Research Institute, represents a doubling of the population every five years for the past couple of decades.

Animals spend the best part of the year as nomads at sea, coming ashore only to breed and to moult. Where they go is uncertain, although feeding grounds have been found along the coast of British Columbia. Here they dive to great depths to feed on skates, rays, small sharks and squid.

Researchers at the University of California at Santa Cruz, headed by leading marine mammal expert Burney Le Boeuf, have attached special recorders to two female elephant seals and have been able to monitor their pattern of dives. One seal broke the seal deep-diving record by reaching 630m (2,067ft). It turned out that they made more than sixty dives a day, staying down for about twenty minutes at a time, and resting at the surface for only three or four minutes between dives. They also, it seems, continue diving both day and night, and it is calculated that they spend about 90 per cent of their lives under the sea. They sleep, it is thought, by 'cat-napping' as they drift down at an estimated rate of 10–20m (33–66ft) per minute. The advantage of spending so much time in the deep ocean, apart from ample time to gather food, is to avoid predation by surface-living sharks.

Each year, at Nuevo, elephant seals start to arrive at the breeding beaches in December. The males land first, emerging from the water to engage in ferocious and bloody battles for dominance of the beach. They warm up with a vocal confrontation. The proboscis is enlarged by a combination of blood pressure and muscular action, and it acts as a kind of resonating chamber to increase the effectiveness of the roar. Individuals square up to each other, chest to chest, and fling their head and blubbery neck at the opponent. Deep wounds are made with short tusky teeth that slash down on soft unprotected flanks, but there are rarely any fatalities. A hierarchy is established which is only maintained by frequent battles. Throughout this period the bulls fast. Only the fittest stay on the beach. Challengers frequently attempt to abduct cows on the periphery of the harem.

Young males show signs of a proboscis hood at about eighteen months, but it is not until eight years later, when the animal reaches full maturity, that it is fully formed. Until then, young bulls have practice bouts, usually without contact. They raise themselves into the threat posture, with necks and heads back and snouts inflated, and make an enormous amount of noise. One pretender retreats before blood is spilled. There will be another sixty or so juvenile fights before the real thing.

Bellowing, it seems, is an important factor in the maintenance of social order. In playback experiments, Jacques Cousteau and his *Calypso* colleagues presented an isolated bull with threat sounds. Startled, it raised itself up, looked about for a rival, and, seeing none, flopped back on to the sand. Similar playbacks to a pair of adjacent, but docile, bulls

Silverback male mountain gorilla dwarfs one of the wardens in the Parc National des Volcans in Rwanda (*Ian Redmond*)

Silverback male eats *Galium* vine (*Ian Redmond*)

**Nest collectors on precarious bamboo ladders in
the Black Cave of the cave swiftlets** (*Phil Chapman*)

Cave swiftlet nestlings in their valuable cup-shaped nest perched high near the roof of the Goodluck Cave in Sarawak (*Phil Chapman*)

Cave racer snake catches a cave swiftlet in mid-air (*Phil Chapman*)

Head of a two metres long Salvadori's monitor (*Ian Redmond*)

Most of the lizard's length is in its tail (*Ian Redmond*)

resulted in them going at each other 'hammer and tongs', each thinking that the other was responsible for the challenge.

Males arriving too early, those eliminated in preliminary bouts, and the old and infirm, haul out higher into the dunes or retire to 'losers' beach' away from the main colony. They spend most of the season in repose, oblivious to the frantic activity around them.

In late December the heads of the females are seen bobbing in the surf: they emerge from the sea, very pregnant, slither up the beach and gather into harems, containing between two and 1,000 cows. Six days later, and usually at night, they give birth to their pups, each blind at birth and clad in its black woolly romping suit. Females usually have one pup which was conceived during the previous year's matings.

As with most seals and sealions, elephant seal milk is so rich in nutrients and butterfat that each pup, weighing about 40kg (88lb) at birth, reaches over 136kg (300lb) in just four weeks. About a third of the colony's pups will die during the first month having become separated, orphaned, or squashed by fighting males.

Soon after giving birth, the cows come into season and are mounted by the dominant males. In fact, they may have already been mounted on several previous occasions, before they are receptive, for the bulls attempt to mate with any female, no matter what their physiological state. Beachmasters mate with as many cows in their harem as they are able, which means that 90 per cent of the females at Ano Nuevo are served by only four per cent of the mature bulls.

After fertilization, the egg does not implant immediately in the uterine wall but is delayed for three months. This ensures that the pregnant female, with a normal gestation period of about eight months, does not give birth at sea, where the young would be unable to survive, but is ready to pup on her return to the breeding beach the following year.

The pup is weaned simply by desertion: the cows return to the sea but do not venture far from the shore. They feed on invertebrates on the sandy bottom, and indulge in the curious habit of stone-eating, a behaviour thought to help achieve the right buoyancy for life at sea. Eventually they will head for deeper waters, returning to the same beach later in the summer in order to moult. Interestingly, mature and immature seals moult at slightly different times, this natural time-sharing process keeping the various age groups apart.

By mid-March, the adults have left the area. The youngsters, now known as 'weaners', gather into small groups and moult their black baby fur, replacing it with a sleek shiny silver coat. By the end of April, the pups head out to sea, possibly in search of the adults in the feeding grounds. Before they go, however, they must learn to swim, and this is where the trouble starts, because young elephant seals are good eating for the killer whale and the great white shark.

The successful recovery of the northern elephant seal populations, together with a similar increase in numbers of harbour seals, Californian and Steller's sealions, and marine otters, have brought with them unexpected problems, and a major conflict arose between conservationists along the West Coast and those who use the magnificent Pacific beaches for recreational purposes. The increase in prey species has been followed by a similar increase in predators, and the great white shark in particular has had some difficulty, it seems, in recognizing its natural prey.

From a shark's point of view, suggests John McCosker, director of the Steinhart Aquarium in San Francisco, surfers on their surf boards look very much like seals. Surf boards, during the past few years, have got shorter and have fins and split tails. Add to this the dangling and splashing arms and legs of the surfer and to a hungry or inquisitive great white the configuration looks a lot like a seal behaving in an abnormal manner – the perfect target.

Shark attacks were virtually unknown along the northern Californian and Oregon coast until 1972. Since then there have been at least twelve. Scientists at the Point Reyes Bird Observatory noted that in the early 1970s there had been no more than one or two shark sightings a year around the Farallon Islands, twenty miles out to sea from San Francisco, but in 1982 there had been seventeen.

The great white shark is a formidable killer. Mature individuals may reach lengths of 4.5–6m (15–20ft) and weigh 900–1,800kg (2,000–4,000lb). A large one could swallow a young 227kg (500lb) elephant seal whole, or take a considerable chunk from an unfortunate surfer. One shark, harpooned off the Californian coast and later examined by McCosker, carried two adult harbour seals squashed into its throat and stomach. Another specimen, taken in the same area, was digesting the rear quarter of an adult elephant seal.

Sharks tend to attack from below and from behind, aiming to take a seal by surprise, slamming into its midriff before the victim has time to escape. An enormous bite taken out of the surf board of a champion surfer, killed on Christmas Eve 1981 off Monterey by a 6m (20ft) great white, is witness to that. Many survivors tell of large sharks hitting them very hard from below and lifting them bodily out of the water. And many *do* survive. After a mouthful of surf board or neoprene rubber suit great whites will sometimes spit out their prey, giving the surfer a chance to get away, that is if they have not received crippling wounds such as a severed artery. There is one theory suggesting that great whites deliberately wait for a victim to bleed to death before going in to finish off the meal.

People swimming or surfing too near to seal rookeries are most at risk, but here there is a further problem. The seal populations are increasing so rapidly that they are beginning to seek new breeding areas, and the only beaches available are likely to be those used by people for swimming, surfing, sailing and scuba diving. 'If they continue to increase on the mainland beaches as they have for the last ten years', says Burney Le Boeuf, 'by the year 2000 you're going to have two hundred and fifty thousand elephant seals, and it's quite possible they'll be competing with people for beach space.'

Growth rates of seal populations must, at some stage, decline. Natural factors, such as the disruption of social structures, the breaking up of harems, delayed maturity, and the presence of a greater number of non-breeding mature females, might limit population growth. Over-crowding on the main beaches is likely to result in pups being squashed or separated from their mothers more often. Wandering pups are mercilessly attacked by other cows. Sometimes foster mothers will take on and feed abandoned pups, but in most cases an isolated pup is destined to die. Younger, less experienced mothers lose their pups more often, and in competition with older females that are higher in the social order, they are forced to move to less crowded beaches. Those giving birth at more exposed sites may lose pups during the fierce Pacific winter storms. In 1978, for instance, when storm surf and high tides occurred in January and February, the peak time for cows and pups, 40 per cent of the 900 pups on Ano Nuevo Island were washed away and drowned.

As yet the optimum population size for northern elephant seals along the West Coast is unknown. Le Boeuf thought that the Ano Nuevo site, for instance, would have reached its peak carrying capacity in 1977, but the population has been increasing at the same rate as before.

Occasionally there are signs of unrest. Researcher Robert Gisiner, working with the seals of Ano Nuevo, found a bull that had been murdered, an event not seen before. Gisiner was chased away by one of the harem, but he later witnessed an unusual piece of behaviour: another bull moved in, began attacking the inert body, and, in what seemed to be complete frustration, attempted to mate with it. It seemed to Gisiner that the animal's instincts had gone completely haywire.

There is, however, a more insidious problem facing elephant seals. They have, since the late 1900s, reproduced successfully, but amongst themselves. The 120,000 animals alive today have derived from the 50–100 that survived on Guadalupe. Each seal's genes are virtually identical to those of the original survivors. The entire population has lost its genetic diversity and is therefore vulnerable to environmental changes. Any adverse natural event, such as a change in climate, a new disease or parasite, that is able to kill one seal is likely to kill the lot.

Inbreeding or mating with close relatives means that northern elephant seal parents pass on a limited range of hereditary information, and the pups, when they mature, in their turn pass on the same information to their offspring. It is known as a 'genetic bottleneck'.

In normal populations, a mixture of genes, half from the mother and half from the unrelated and genetically different father, are given to the offspring, allowing 'rare' and potentially useful genes to be present in the genetic make-up. If the 'rare' gene programmes the animal to withstand a new disease, then that animal will survive to pass on its genes. All those without the 'rare' gene will die. Subsequent generations will thus gain the 'rare' gene and the resistance to the disease from the survivor. Inbreeding rules out the likelihood of 'rare' genes appearing because the same genes are passing from generation to generation throughout the entire population. The future survival of the Pacific population is not ensured by natural selection for the northern elephant seal lacks this built-in adaptability to environmental change.

11

The Americas:

ii. South and Central America

South America is dominated by two great geographical features – the longest mountain chain in the world, the Andes, and one of the greatest river systems, the Amazon Basin. There are also several other major regions – the grasslands of the Pampas, the seasonal wet and dry lowlands of the Grand Chaco, the marshy Pantanal, the Atacama Desert on the Pacific coast, where, it is said, no rain has fallen for over 400 years, the sub-Antarctic islands of Tierra del Fuego, and the wind-swept steppe of Patagonia.

The continent was once joined with the Antarctic and Australasia, and still today marsupial animals, such as the woolly opossum, the lutrine and the monito del monte – all American opossums – live here and in Central America, and southeastern North America.

Central America is sandwiched between the Pacific Ocean and the Gulf of Mexico, and is remarkable in that the two expanses of water exert different influences on the two coasts. Climate, vegetation and animals on either side of the dividing high ground are quite distinct – jungle and cloud-forest on the Gulf side, and grassland and open forests on the Pacific side. The region acts as a gigantic funnel for bird migrants moving to and from North and South America.

An Extraordinary Bird

In the swamplands and along the river banks of the Amazon and Orinoco there lives a bird that swims before it can fly, flies like a bloated chicken, eats green leaves, has the stomach of a cow and has claws on its wings when young. It would not be out of place in Lewis Carroll's *Alice in Wonderland* or amongst the fantastic animals of Edward Lear, but it is real and was once described in 1918 by William Beebe as 'the most remarkable and interesting bird living on the earth today'. It is called the hoatzin.

In appearance, the adult looks like a cross between a domestic chicken and a secretary bird. Male and female look very much alike with brown on the back and cream and rusty red underneath. The head is small, with a large crest on the top, bright red eyes, and electric blue skin. Its nearest relatives, according to biochemical analyses of DNA (the hereditary material contained in each of the bird's cells), are the cuckoos. Its most remarkable feature, though, is not found in the adult but in the young.

Fledgeling hoatzins have a claw on the leading edge of each wing, at the point of the thumb, and another at the end of each wing tip. Using these four claws, together with the beak, they can clamber about in the undergrowth, looking very much like primitive birds must have done. *Archaeopteryx*, which lived about 140 million years ago and is one of the earliest known feathered birds, had three claws on each wing. The hoatzin, however, could not be considered primitive. It is a highly specialized bird.

During the drier months between December and March hoatzins fly about the forest in flocks containing 20–30 birds, but in April, when the rainy season begins, they collect together into smaller breeding units of 2–7 individuals. They build their nests over water and there is considerable competition for the best nest sites. Each breeding group defends a strip of river bank territory, about 46m (150ft) long, against the intrusion of other hoatzin groups that are attempting to improve their lot. Island sites are particularly at a premium because they are less accessible to predators. The nest itself is an untidy platform of twigs built about 4.6m (15ft) above the river, an important feature for the survival of the young.

When danger threatens, in the form of a snake or a monkey, the young hoatzins – maybe three in one nest – dive over the side and into the river. They swim about under the water until it is safe to return and then, using their claws, haul themselves up through the branches and back into the nest. When they have learned to fly they lose their claws and escape predators, not by swimming but by flapping off, in a rather ungainly fashion, to a neighbouring tree.

Unlike most other birds but in common with a hundred or so species of 'cooperative breeders', it is not just the parents that raise the new brood, but also 'helpers'. These are usually related to the parents, perhaps juveniles hatched the previous year, and they take part in all aspects of bringing up this year's youngsters. They help to build and maintain the nest, feed and incubate the chicks, and defend the territory; it is as if they were practising for the future. Stuart Strahl, of the New York Zoological Society, who has been studying the hoatzin in Venezuela, has discovered that about 60 per cent of the birds in this area have helpers, and they raise 50 per cent more young than those without. Predators account for over half the chicks in any one year. The hoatzin nest is particularly vulnerable to predation by the black-capped capuchin which can locate nest sites simply by the pungent smell. This is the result of the peculiar diet of the birds. They eat green leaves.

Stuart Strahl's hoatzins like *Margaritaria* leaves which they forage early in the morning. They may flutter up to 152m (500ft) away from the nest in order to collect the choicest morsels. By eating leaves, though, they have a digestion problem and they have solved it in the same way as other herbivores. The crop has become enlarged and acts like a cow's rumen. Bacteria in the crop break down the cellulose, which releases the nutrients. With such a slow digestion process and low-energy food, the bird cannot be a frequent flyer, for it does not have the energy readily available. Also, its flying muscles have been considerably reduced to provide space for the enlarged crop. It can, however, fly for several hundred metres without tiring.

Egg collecting, hunting and habitat destruction are the main long-term threats to the hoatzin. In Surinam and French Guiana they have become very rare, although in other parts of South America they are still abundant.

Tickle on the Chin

The manakins are small colourful birds, the size of tits, that live in the tropical lowland forests of South America. The females of most of the fifty known species are less flamboyant than the males, which are often black and white with splashes of red, orange, and blue. Courtship is often a noisy and colourful affair.

The male white-bearded manakin, for example, performs in a lek. The arena has a single vertical sapling – the main display perch – and several subsidiary perches. All leaves and twigs that can be carried are removed. The display consists of a series of jumps and gyrations, accompanied by snapping and whirring sounds from modified wing feathers. The female dances with a male and, if he is found to be satisfactory, mating takes place on the display perch.

Some species display on a horizontal perch, and include rapid side-steps, turns and fanning of the tail in their repertoire. One manakin species turns its body rapidly from side to side and makes a grasshopper chirring sound, and the entire display is performed upside down. Another variation is to crouch on the ground and approach the female with wings spread.

The most dramatic, perhaps, is the 'catherine wheel' dance performed jointly by three male swallow-tailed manakins. When the female arrives on the display perch the first male approaches her, leaps into the air, hovers, and then flies backwards, in a flurry of twangs and whirrs, to the back of the queue. As he lands the second bird follows suit, and then the third, until they are performing in a seemingly endless chain. Only one male, the dominant one, mates with the female.

The male wire-tailed manakin is sometimes joined by another 'support' male and they perform an exquisite *pas de deux* in order to entice the female closer. When she arrives on the display perch, the lead male has a unique display that has only recently been carefully observed. The male of this species has some of its tail feathers modified into downwardly curved filaments, much like a splayed-out wire brush. At first it was thought that the wire tail was simply an additional visual element in the bird's display, but now it is known, from the work of David Snow at the British Museum (Natural History) at Tring and Venezuelan ornithologist Paul Shwartz, that the modified tail structure is a tactile organ.

The male wire-tailed manakin backs up to the female, raises its tail, and wiggles it from side to side just under the female's chin. This tickling display appears to establish a temporary bond between the pair prior to mating. Experienced females will even place themselves at the right height and in the correct position on the perch to get tickled.

Univited Dinner Guests

Tropical orb-weaving spiders of the genus *Nephila* construct the largest spider webs in the world. The silk is said to have the tensile strength of steel but the elasticity of a rubber-band. Travellers have told tales of running into a web and literally bouncing out. It can be 2m (6ft) in diameter and have guy lines extending 6m (18ft) into the trees.

The spiders are successful at trapping a good supply of flying insects, many of which cannot be eaten immediately and are wrapped and stored in silk cocoons. This has not been lost on passing opportunists, and huming birds, warblers, scorpionflies and wasps steal from the web. There are even miniature flies that perch on the spider and take small titbits from the meal being eaten. But, perhaps, the most audacious kleptoparasite on the web of the golden silk spider *Nephila clavipes* of Panama is a very small silver-bodied spider which belongs to the genus *Argyrodes*.

The association has been studied by Fritz Vollrath, at the Smithsonian Tropical Research Institute, and he has found that not just one spider lies in wait to steal from the host, but up to 45. The little spiders wait around the periphery of the web with thin signal threads attached to the hub area where the host keeps its larder of silk-wrapped prey. When the vibrations indicate that the larger spider is unwrapping a meal, the parasites rapidly run out. They take care to move only when the host is active or they may end up on the menu themselves. When danger looms large they simply drop out of the web suspended on a dragline.

If the host has hold of the food item, the small spider simply feeds on the opposite side. If, however, the prey is loose, the parasite may steal it from right under the spider's mouth. Anchor lines are attached to the food cocoon and fastened to a safe point outside the web. The main web is cut and the food swings away to be consumed at a safe distance. When too many parasitic spiders take over the web the host moves out and builds a new web elsewhere.

One of the inherent problems of giant webs is the risk of bird collisions. *Nephila* minimizes the danger by having fairly obvious yellow and grey silk. Other large orb-web spiders have devised an ingenious way of weaving extra silk into the fabric of the web that acts as a traffic sign. *Argiopes aurantia*, observed by Tom Eisner and Stephen Nowicki, of Cornell University, contructs a broad band of silk from the top to the bottom of the web, known as a stabilimentum. The name arises from the fact that it was thought previously that the structure is built-in to give the web extra strength. The researchers found that after dawn only 8 per cent of unmarked webs remained intact whereas over 60 per cent of those with stabilimenta survived.

The Frog and the Bat

The mud-puddle frog of Barro Colorado Island's rain forests is a noisy little animal but, unlike other frogs and toads, it has an attractive – even musical – mating call that is divided into two parts. There is a soft 'aow' and a hard 'chuck', and between sunset and sunrise a male mud-puddle will emit up to 7,000 of them in an attempt to attract a female. Females prefer males giving both parts of the call – the 'aow' is the turn-on and the 'chuck' is a sound that makes the male easier to locate. Also, the deeper the pitch of the 'chuck' the more attractive a male became. Males, though, did not always give the full call. They would often leave off the 'chuck'. Stanley Rand of the Smithsonian Institution and Michael Ryan of Cornell University wondered why the male does not give the full call all of the time? Surely a male is not going to give less than his best in order to attract a mate? The solution to their puzzle swooped down from the sky – not a heron or a hawk, but a frog-eating bat.

The predator that caused the frogs to pipe down was discovered by Merlin Tuttle, of the Milwaukee Public Museum. One day on Barro Colorado he found a bat with a frog in its mouth. It was the fringe-lipped bat and the victim was the mud-puddle frog.

The researchers carried out a series of experiments in the forest and they found that the bats would even swoop down to attack a loudspeaker playing the frog calls. They were not using echo-location to home in on their targets but were detecting and locating the frogs by the sounds that they made. They could also discriminate the calls of edible frogs from those of poisonous or dangerous ones. One large toad has been seen to turn the tables on an attacking bat and grab it as it flies down. The bats understandably ignored the calls of toads.

The bats, though, do not have things all their own way. The mud-puddle frogs can also detect the presence of the bats, but only on nights when the moon is visible. They use visual cues as an early-warning system, and take evasive action by stopping their calling.

Fish and Nuts

The removal of vast tracts of tropical rain forest not only destroys potentially useful plants, for food and medicine, and untold numbers of undiscovered terrestrial animals, but also potentially commercially useful freshwater fish. Research in the Amazon has revealed that there is often a very close relationship between the fish in the rivers and the trees on the bank.

Piranhas have an evil reputation. A large shoal, it is claimed, can demolish the carcass of a dead cow to its bare bones in just a few minutes. There is an instance recorded of a 45kg (100lb) capybara being stripped in less than a minute, and there is the story of a man and his horse falling into the water and, when their skeletons were later recovered, the man's clothes were still intact. Blood in the water is thought to trigger a feeding frenzy, but they do not always attack. Local peoples often bathe and swim in water known to contain piranhas and they emerge quite unscathed. Others report that they have been attacked whilst wading across a shallow ford. In some regions villagers have toes or fingers missing. Women are particularly at risk when washing clothes.

The bite size of the piranha is relatively small. With small, sharp, triangular teeth it can take a 16cu. cm (1cu. in.) chunk of flesh, although many of its normal prey species are swallowed whole. The surprise comes, however, when we discover that piranhas are also fruit and nut cases.

A shoal of piranhas has been seen waiting below the branches of a rubber tree for the pods to pop and the seeds to fall. Many do not even reach the ground as they are eagerly gobbled up. The odd thing is that the event was taking place 48km (30 miles) from the nearest river. How did the piranhas arrive there?

Every rainy season, between June and November, the Amazon and its tributaries flood about 100,000 sq. km (40,000 sq. miles) of the surrounding forest, and the piranhas follow the flood water. They become a significant part of the forest ecosystem, feeding on fruits and nuts, and distributing the seeds far from the parent plant. Other species of fish follow the same migration and have become well adapted for fruit and nut eating, with large crushing molar teeth and the propensity to put on fat to survive the period when the waters recede. Researchers at the Instituto de Pesquisas da Amazonia have found that several species of piranhas have given up meat-eating altogether and have become vegetarians. In all, about 200 species of trees and fish have been found to rely on this interaction.

The important factor to emerge from the study, however, involves man, for local Amazon basin communities rely heavily on fish from the rivers for their main supply of protein. Devastate the forest and the vital protein source is lost too.

The Black Cayman

About 99 per cent of all the black cayman that were living about a century ago have been wiped out by hunters. Little is known about their biology or behaviour for they were brought almost to the point of extinction before anybody bothered with them at all. Today, only a few survive in remote lakes in the tropical forest areas of Peru and neighbouring countries. They are now fully protected.

It was not the local peoples that massacred the black cayman. They have a great respect for this creature that can reach a size of 6m (20ft), and it is an important animal in their folklore. They do take a few for medicinal reasons – the fat is supposed to cure chest complaints. Much of the slaughter was in the 1950s and 1960s when helicopter gunships would go to remote lakes at night and shine powerful searchlights on to the water surface. The cayman could be spotted by the light reflected back from their eyes and then they would be shot. The following day another team would go to the lake and collect the corpses.

A group of researchers concerned with the conservation of these creatures has addressed itself to the fact that cayman may be living in one lake and yet be absent from an almost identical lake nearby. When the cayman population of a lake becomes greater than it can support, young members of it, usually about 1.5m (5ft) in length, disperse to other lakes. Local villagers reported that, years ago, the lakes without cayman used to contain large numbers but that they have not been repopulated.

A factor found to be important is that the empty lakes have a larger than average population of piranha fish. With the absence of mature cayman to keep them in check, the piranhas have increased their numbers enormously, and these voracious predators are patrolling the lakes biting off the legs and tails of any young cayman they find. At the size at which they disperse, the cayman are still vulnerable to the attacks of piranhas, and so are unable to colonize the new lake.

A census of cayman living in the study area has revealed that the larger specimens live in the marshes and the smaller ones frequent the main body of the lake, coming out by night into a 4m (13ft) square territory. In October, when the rainy season begins, the 1.5m youngsters take advantage of the floods and disperse, in attempts to find new homes and to stock other lakes.

The research team used radio tracking to find where the cayman spend their day, and discovered that the young black cayman burrow in the mud. Here the young ones are safe from predators, and the temperature is a constant 22°C (72°F) – a reasonable temperature for a cayman to be at, without wasting too much energy, and not too cold for it to be able to react quickly if threatened. The larger specimens are thought to bask in the marshes, but not out in the open, where they would be vulnerable to hunters' guns. The behaviour of black cayman seems to be modified to avoid any encounter with man.

Turtle Tears

Turtle Mountain is a 213m (700ft) high block of volcanic rock that pushes out into the Caribbean Sea. Below it sweeps the white arc of Turtle Bogue, a section of Costa Rican coast famous for the annual arrival of green sea turtles. The turtles migrate from their feeding grounds around the Yucatan Peninsular and the Venezuelan coast and somehow navigate to the Bogue to lay their eggs.

They arrive several times a year, and they come in their hundreds. During the day, flotillas of bobbing heads appear in the surf. Nobody sees them coming, they just suddenly appear. Most are females, but dotted amongst them are one or two males. The males do not leave the water, but simply float about just off the beach waiting for the female of their choice to swim by. They are not gentle. Several males might jostle one female, each one butting this way and that in an excited frenzy. Mating is violent. The male grasps the female with large grappling nails, located on the first digit of the front flippers, and clambers on to her back. The strong horny tip of his prehensile tail wraps around the rear of her abdomen. The three-point coupling gives the pair some measure of stability in the surf, and the grasp can be so powerful that a male can rip into the bones and shell of the female. After copulation the sperm is stored, to be used in future seasons. It has been suggested that these sperm will serve to fertilize eggs for the next laying in three years' time. With mating complete, the male returns to the deep, as mysteriously as he arrived, and the female begins a difficult and sometimes dangerous journey, out of the sea and up on to the beach to lay her eggs.

It is, perhaps, an evolutionary irony that the female sea turtle makes this terrestrial excursion. The reptiles had successfully colonized the land and had become independent of the sea by developing the soft-shelled egg – virtually an aquatic-marine environment in miniature. Despite having returned to live in the sea, turtles must visit the land in order to deposit their capsules of seawater, inside of which the new baby turtles will grow.

Bleeding and with half-inch wounds cut into the bone, the female green turtle rides gently in on the waves until she reaches the beach and runs aground on the wet sand. It is night, and the waves break over her with an iridescent spray. Her dark shell glistens in the moonlight. She sits for a few moments, her huge primeval head swinging from side to side as if checking her landfall. She suddenly drops her head and rubs her nose in the sand; perhaps an olfactory clue tells her that she is in the right place. The slightest noise will frighten her away.

If undisturbed, she slowly and laboriously hauls herself away from the tideline towards the dry sand at the top of the beach. She breathes with huge sighs. Her whole body, normally supported by the water, presses down on her lungs and windpipe. In the sea, her thoracic cavity expands ventrally, but now she must lift her entire body away from the ground in order to fill her lungs. She coughs, grunts and sighs continuously.

Pulling with her front flippers, she drags herself inch by inch across the sand. Glutinous tears hang down below her eyes, the mucous not only involved in salt regulation but also keeping her eyes clear of particles of sand. She stops often, peering around in the gloom, and eventually reaches the place where the sand is light and of medium-coarse texture. It must not pack too tightly or bake into a hard surface.

First she digs a shallow pit which will conceal her during the vulnerable egg-laying period. She grunts, and sweeps sand to each side with her front flippers to make a 46cm (18in.) deep depression. The front flippers slow down and the scooping movement of the hind limbs means that she is beginning to excavate the egg chamber. It is dug to a depth where temperature and moisture will remain roughly constant. She sighs deeply as she pushes the edge of the left hind flipper into the sand below her. She picks up a small amount and deposits it to one side. The right hind flipper, until now steadying the body, jerks forward, kicking away any loose sand.

Up to the point of egg-laying the turtle will abandon digging it disturbed and return to the sea. When egg-laying begins she seems to go into some kind of trance and nothing will deter her. Wild dogs have been known to take the eggs as they are being laid. The female shields the bottle-shaped egg chamber with her hind flippers and lays about a hundred soft spherical eggs. They drop into the pit in twos and threes.

With egg-laying complete, the hind flippers scrape and push the sand about until the pit is filled and a small mound of sand is left on the surface. The female kneads the mound by a short chopping action with the inside of the front flippers. She fills the body pit and scatters sand around in order to disguise the site. She turns and heads back to the sea, leaving a track of disturbed sand extending from the nest site. On nearing the sea, she quickens her pace. The small waves splash over her, and suddenly she is gone, heading back once more to the feeding grounds, maybe thousands of miles away. The entire process has taken about two hours, and this is the fourth time she has undertaken the journey across the beach. In all, this season, she has laid over 400 eggs.

Egg-laying, for the new generation of sea turtles, marks the start of a hazardous journey through life that makes one wonder how the creatures survive at all. If the eggs are not eaten by dogs before they can be buried beneath the sand, then there is still the chance that they will be dug up by buzzards, peccaries, coyotes, opossums or a host of other beach scavengers that suddenly appear on the Bogue at nesting time. The first two days after laying is the most dangerous period for the eggs. After that, the beach animals are unable to find the nests and the sixty days of incubation, under the sand, follow in relative safety.

The newly-laid eggs are not completely filled. They absorb water from the damp sand and increase in weight. Each fertile egg shows a white spot at the upper pole just 24 hours after laying. At this point, the albumen separates from the shell and in the centre the embryo develops. After two days the embryo consists of five body divisions; at ten days the heart forms and is beating rapidly and regularly, the flipper buds grow, and the eyes are well advanced; after 30 days the ribs and plates can be seen and the eyes appear very big; at 38 days the carapace becomes coloured, and the head, lower jaw and flippers are able to move freely. During the development the calcium and magnesium required to build the new turtle is five times that found in the yolk. The additional material is taken from the egg-shell, which contains an outer layer of lime.

Nine or ten weeks after egg-laying the baby turtles hatch out. The first to break out of their shells do not start to dig their way to the surface immediately but wait for their nest mates to join them. As more and more emerge, the hatchlings thrash about wildly. The tiny turtles at the top of the nest scrape away at the ceiling, while those at the sides undercut the walls, and the ones at the bottom trample and compact the sand filtering down from above. The cooperation is unintentional but it will allow them all to reach the surface. Sometimes the activity stops, and it is up to the bottom shift to squirm, and trigger another period of movement. Thus, in a series of fits and starts, the ceiling falls, the floor rises, and the chamberful of hatchlings gradually erupts on to the surface. Without this cooperation, the ones on the lower levels would be unable to fight their way to the surface.

Once on the beach, the bonds between nest-mates loosen, but even so a batch of hatchlings moves faster across the beach than an isolated individual. Straying or lethargic turtles are put back on the straight and narrow by the rush of the others. After a few false starts, most youngsters point accurately towards the sea, and can orientate in any weather, during the day or night, and with or without cloud cover. The quality of light over the sea has been suggested as a directional clue. They do not head for the moon or sun, and become disorientated if blindfolded. They must get quickly from the nest to the sea, for this short journey, perhaps as little as a hundred yards, accounts for the greatest losses. Ahead, on the sand, lies danger after danger.

Green turtle hatchlings rarely emerge during the daytime. The warmth of the sun seems to inhibit any emergence. At night, they may avoid the monsters of the beach. Those that do make a run, when it is still light, attract the attention of hungry predators. The trees begin to fill with vultures, crows and buzzards. As the exodus starts, the birds swoop down, tearing the heads and flippers off the little turtles. They gouge out the fleshy parts leaving the shells scattered like tiny tombstones about the beach.

The birds cause the walls of an egg-chamber to collapse, making it impossible for the remaining hatchlings to escape. They are picked off one by one. Surrounded by such a sudden abundance of food, even the most voracious vulture can be quickly sated, but nevertheless the young are still dragged out and tossed aside. Some get away only to be intercepted by a foraging iguana, a huge prehistoric monster-like lizard that grabs the hatchlings in its powerful jaws and crushes them to pulp, swallowing them in one or two gulps.

In the next zone are the ghost crabs. Out of their holes they come, grabbing baby turtles with their huge claws. If they catch the rounded shell the baby may be able to wriggle free, but a firm pincer on a flipper means the end of the turtle. It is dragged to the crab's burrow and devoured at leisure. The crabs are able to move very fast on the sand, scanning the beach with their periscope eyes, but they will not chase a turtle far from the burrow for they are wary of trespassing in another crab's territory.

Some hatchlings make it past the crabs, but on reaching the tideline another predator drops in from the sky. In an incredible display of aerobatics, frigate birds scoop up the turtles, swallowing them whole while still on the wing. For those that make it to the sea the dangers are not over. A black triangular fin, cutting through the water just beyond the surf, marks the presence of a patrolling shark.

Curiously, the hatchlings begin their swimming strokes on the wet sand at the water's edge and, without practice, they are able to enter the waves and swim rapidly away, following the same compass heading as they were taking on the beach. They swim about 20cm (8in.) below the surface, coming up for a gulp of air every five to ten seconds. Small fish are able to catch them, but when an aerial predator flies over, they crash-dive down to a depth of 3m (10ft). To nourish them on the outward journey they carry a yolk sac, a left-over from the egg. They have been known to swim as much as 80km (50 miles) from the breeding beach to find a raft of sargassum, their first drifting home, where they are afforded protection amongst the fronds and supplied with food. Occasionally, they will swim from one patch of weed to another. The rafts can be as small as 10cm (4in.) across or as much as 100km (60 miles) long. The young turtles are predators themselves at first. Tiny turtles are weaned on jelly-fish, shrimps, and other crustaceans that also hide in the weed. Unfortunately, other materials, particularly man-made pollutants, also accumulate in the same places as the sargassum and little blobs of tar, discarded by the petrochemical industry, gum up the jaws of many inquisitive baby turtles causing many deaths.

The survivors may travel for thousands of kilometres in the ocean currents. The question is, could they be drifting towards turtle nurseries? This next stage in the turtle's life story is still a mystery, for the youngsters are not seen again until they reach the size of dinner plates, and they become herbivores, feeding on eel-grass in sheltered coastal bays. They do not become mature until they reach a weight of about 90kg (200lb). Then, they return to the beach on which they were hatched to start the whole process off again.

12
The East

The East is influenced by two main expanses of water – the Indian and Pacific Oceans. There is a major young, mountain chain – the Himalayas – containing some of the highest mountains in the world. There are deserts, such as the Gobi Desert of China, tropical jungles in southeast Asia, wetlands such as the Sundarbans, and volcanic island chains such as Indonesia, the Philippines and Japan.

Here, east meets west and north meets south. It is a crossroads of animal movement. On the Indian sub-continent, for example, jackals, lions, elephants and gazelles from Africa have met civets, dogs, rhinos, and the ancestors of baboons and guenons from Malaysia and Indo-China, along with wild boars moving southeast from Eurasia, and tigers heading southwest from Siberia.

Tiger, Tiger

The population of tigers in India was almost at the point of extinction in the early 1970s, but today, thanks to Operation Tiger – launched by the Indian Government in 1973, and supported by the World Wildlife Fund and the International Union for the Conservation of Nature and Natural Resources – it is recovering, having doubled its numbers in thirty years to over 4,000 individuals. It still has a long way to go before reaching the estimated 40,000 tigers that lived at the turn of the century, but then again, there would not be the space for them. Indeed, there are already problems of tigers, with insufficient home range and prey, turning to man-eating to supplement their diet. In the first three months of 1985, for example, 22 people were killed in the forests of West Bengal despite attempts at dissuading tigers from taking humans by conditioning them with electrified human dummies.

A similar situation was experienced in the Sundarbans – an area of mangrove forest, water channels and islands on the delta of the Ganges and Brahmaputra rivers in Bangladesh and India – where fifty people a year are killed by tigers. Here, these big cats ambush fishermen in their boats, and woodcutters and honey collectors on the forested islands. Dummies, wired up to give the pouncing tiger the shock of his life, were placed in boats and in forest clearings, and, according to some reports, there was some measure of success.

The tiger is an elusive, solitary animal, and so very little was known about its biology until formal studies were undertaken at the Royal Chitawan National Park in Nepal. Using radio-tracking techniques (a radio-transmitting collar is attached to an individual, and it is followed by a scientist with a directional receiver), researchers have found that a male has a large home range, up to 100sq. km (40sq. miles), encompassing the smaller home ranges, 20sq. km (8sq. miles), of several females. The home ranges of animals of the same sex do not run into each other at Chitawan, although elsewhere female ranges have some degree of overlap. Boundaries have to be constantly patrolled, for if an animal does not visit a corner for a month or so, it will be taken over by its neighbour. Scent marks, from a mixture of urine and anal gland secretion sprayed on bushes and rocks, scraped faeces on paths, and scratched trees are used as 'keep out' signs.

The scents left by a female may also indicate to the male her readiness to mate, an important signal as females are only receptive for a few days at a time. When the female is ready to give birth she finds a safe and secluded 'den' site. The cubs, usually three or four in a litter, stay here for about eight weeks, after which they follow mother everywhere for they depend on her for their daily food until they are about 18 months old. They are permitted to stay in their mother's range until they are about 2½ years, and then they must leave to find a home of their own.

A hunting tiger detects suitable prey animals by smell and will either stalk it or wait in ambush. When it is close enough, the tiger will rush forward and jump on to the victim's back, ensuring that its back paws are on the ground to topple the prey over, and clamp its jaws around the neck. The prey is dragged into cover and, if it is large – sambar, baby elephant or rhino – it is partially consumed, rump first, and then visited again until only the bones are left. Small prey animals such as barking deer or wild pigs are demolished in one sitting. A female with cubs needs to make a kill every five days, which is no mean task if you consider that it sometimes takes twenty tries before a successful kill is made.

Tigers have had a tough time since the middle of the nineteenth century. Prior to that, there were many sub-species distributed throughout Asia – the Indian tiger, the darker Indochinese tiger, the reddish South Chinese tiger, the brownish, narrow-striped Caspian tiger, the enormous, long-coated Siberian tiger, the hairy-jowled Sumatran tiger, the thin-striped Javan tiger, and the small Balinese tiger. Today, the Balinese is extinct and the Caspian, and South Chinese populations are thought to be extinct, although there is some speculation that a few Caspian tigers still live in the mountains of Iran. The Siberian tiger has been reduced to about 200 animals, and the Sumatran down to 600. The Sumatran is protected but villagers, fearing for their lives, trap and kill them still. There are thought to be just four Javan tigers capable of breeding that live in a reserve at Meru Betiri, in East Java, although there are also thought to be several isolated animals elsewhere.

It has been estimated that tens of thousands of tigers were shot for sport from 1850 to 1950, with some people, notably British, Indian or Nepalese big-game hunters, taking more than 300 animals each. King George V and a shooting party in 1911 took 39 tigers in eleven days. In 1921, the Prince of Wales shot 17 tigers in Nepal and another seven in Gwalior the following year. In the winter of 1938–9 the prime minister of Nepal and some guests shot 100 tigers in the Chitawan Valley. The Maharaja of Udaipur claimed a 'bag' of over 1,000 tigers and the Maharaja of Surguja, who died in 1958, took the all-time record of 1,707 – more than a third of the tigers living in India today. When an inexperienced guest went hunting, a 'crack' marksman was placed discreetly behind, and he would shoot at the same time as the guest. In this way a host was assured that his guests returned home satisfied that they had killed a tiger.

The Deer and the Monkey

Chitals are small spotted deer that live in the forests of India. When food is plentiful, they graze on grass and low-lying vegetation in a scattered herd, but, in the dry season, when times are hard, they congregate together under a tree in which a troupe of common langurs is sitting.

The langurs are rather wasteful creatures, eating the leaf stalks but discarding the leaf blades. It has been estimated that a single troupe of 20 langurs will throw down about 1.5 tonnes of foliage during a year and about 0.8 tonnes can be utilized by chitals. Forest birds might also add to the bonanza by dropping pieces of fruit.

The association, which can last for as much as eight hours a day, has certain other advantages both for the deer and the monkeys. Deer can detect approaching predators by smell, while the monkeys have keen eyesight. Not surprisingly, chitals and langurs respond to each others' alarm calls, for both can fall prey to tigers and leopards.

The Giant Panda

There are thought to be no more than a thousand giant pandas living in the wild. It is not only one of the rarest animals in the world but also one of the most popular, two factors that led the World Wildlife Fund to adopt the panda as its symbol.

The panda is a relatively new arrival on the scientific scene. It was not discovered and recorded until 1869, and the first captive specimen was brought out of China as recently as 1917. Little is known about the animal in the wild, and it could be that little will ever be known as giant panda numbers dwindle. Most of the research in the field has been carried out by Hu Jinchu and George Schaller, co-leaders of the WWF-China panda project.

'For any species with only a thousand animals, scattered in small isolated populations', says George Schaller, of the New York Zoological Society, 'there is a real danger that they may become extinct in the wild.'

The few giant pandas that have survived now live in remote and inhospitable mountain reserves where they are protected by law. Although these animals are relatively safe from any further human activity, the reserves themselves are isolated and so pandas are unable to move freely between them. This has proved to be a major problem because the bamboo, on which they depend for 99.9 per cent of their food, is gradually dying off. This is not abnormal – the bamboo plant flowers, seeds and dies back every 50–60 years. This leaves the pandas with no food. The bamboo, though, does not die off everywhere at the same time. If the bamboo forests were connected then the pandas could move to areas that were unaffected, and no doubt this is what they did in the distant past. But, today they cannot migrate. In 1975/6 the bamboo of Wangland reserve flowered and 138 pandas died of starvation. In 1983, the bamboo growing in the Wolong reserve in the Qionglai Mountains of Sichuan began to flower and die. About a third of the world's giant pandas lived here, but only twenty dead pandas were reported to have been found.

There are two main types of bamboo favoured by the giant panda – the 2m (6ft) high arrow bamboo growing between 2,500m (8,200ft) and 3,400m (11,150ft), and the 4m (13ft) high umbrella bamboo is found between 1,600m (5,250ft) and 2,600m (8,530ft). In Wolong it was the arrow bamboo that flowered. In the past the pandas would have moved down to the lower slopes and fed on the umbrella bamboo, but today the umbrella thickets have been cleared for agriculture. At Tangjiahe, a reserve with few people and little disturbance, the pandas are feeding on three species of bamboo and so with such a choice are not so badly hit by the flowering. There are, in fact, about twenty-five species of bamboo in panda country, but most of those growing on the lower slopes have been removed. Crises have occurred when several species of bamboo have flowered, over a large area, at the same time. In the main, the bamboo flowers in isolated patches and so the pandas can easily move over to another patch or another species.

Each adult panda requires at least 15–20kg (33–44lb) of bamboo each day. It will sit on its haunches, hold a stalk in its paws, and peel back the outer layer before crunching the centre part. It can hold the shoot steady by using an enlarged wrist bone that acts like a thumb, and crush the stems with flattened molar teeth. During the winter and early spring, the pandas seem to prefer the stems and in July to November concentrate on the leaves. They have been known to take the occasional mouse or juicy insect that they might find in the bamboo thickets. It is thought that the giant panda was once a carnivorous beast, and that it gradually converted to a herbivorous diet as the wildlife in the area disappeared and its ability to catch other animals diminished. In the recent past, it most likely supplemented its diet with carrion, such as musk deer and golden monkey, left over from leopard kills. One question puzzling scientists, though, is why the panda chose to eat bamboo at all. It is not nutritious. Neither is the panda's carnivore-type anatomy and physiology geared to breaking down cellulose and such like. It has a reinforced oesophagus, but a very short gut length, and so it takes about ten hours to process a meal and only digests about 20 per cent of all the food that it eats.

A panda's daily routine starts in the early morning when it begins to search for the best bamboo stands. It will spend the rest of the day in the thicket, coming out again in the evening to resume the search. It averages 14.2 hours in the day simply feeding. When thirsty, it digs a hole by a stream, waits for it to fill with water, drinks as much as it can hold, and waddles away with a distended belly. In the summer it spends its time in the cool of the mountains, where it might eat other plants as well as bamboo, and in winter comes down into the shelter of the valleys where, in the winter snow, only bamboo is available to eat.

The panda is a solitary animal. The male has a larger home range than the female. Scent marking posts scattered throughout the range ensure that pandas avoid each other, except when it is time to mate. The main danger in the home range is likely to be the traps that hunters have set for musk deer. The pandas are accidently trapped and killed.

If it survives the bamboo famine, the traps, and does not succumb to internal parasites (2,000 roundworms were found in one dead panda, and a radio-collared specimen was found to have 172 large parasitic worms, some 14cm (5.5in.) long and blocking the pancreatic ducts), a panda will live for 20 to 30 years.

A male becomes interested in the opposite sex when he reaches his sixth or seventh birthday, while the female matures at about four or five years old. Just once a year, during April and May, a female will indicate her readiness to mate by emitting a series of grunts and bleats, and by rubbing her anal gland against trees and stones. The male responds with his own high-pitched sounds. If he does not reply, he will not be allowed to mate. If he does reply, more often than not the couple will climb a tree to avoid the attentions of rival males and passing predators. Several males might visit a female in the course of the season.

The gestation period can be between 97 and 163 days. The implantation of the embryo is delayed and so, even though there appears to be a long gestation period, the panda cub, in reality, is not that well developed at birth. It is blind and weighs 100–150g (3–5oz). Its post-natal development, though, is rapid. It will open its eyes when it's a month and a half or two months old, move about after three months, and may be independent after a year. The pregnant female seeks out a makeshift den, such as the hollow base of an old fir tree, in which to give birth to a single cub. Sometimes she may have twins or, on rare occasions, triplets. It is unlikely that more than one cub would survive in the wild as the mother must feed the baby every hour for the first week of its life, and that is only possible by attending to one cub at a time. One mystery is that often baby pandas disappear before they are three months old. Nobody knows why. Leopards and jackals are the main predators that take young pandas.

Why pandas should be black and white is not clear, either. The markings might warn predators that they are not to be tangled with, or it could be cryptic colouration for moving about discreetly at dawn and dusk.

Another mystery which has only recently been solved is the family into which giant pandas and red pandas should be put. Work with panda DNA (the blueprint of life contained in every living cell) has revealed that the giant panda is most definitely a bear, and that the red panda is a racoon. The two families apparently split and went their separate ways some 50 million years ago. The giant panda split off from the rest of the bears about 20 million years ago.

Wild Cow of the Forest

In 1982, a Thai zoologist surprised and delighted the conservation world with news that the world's oldest and largest wild cattle are not extinct. He was referring to the kouprey, an animal not seen since before the Indochina War and thought to have been annihilated during the fighting. In fact, it is a wonder any animals have survived in that area at all for three major wars have raged back and forth across the forests – Japan against China, the French against the Viet Minh, and the USA versus the Viet Cong.

The bull kouprey is 2m (6ft) tall at the shoulder, and carries the longest and widest horns of all living cattle, except the water buffalo. It has large dewlaps below the jaw that may reach down to the ground. The tips of the horns tend to split but the split ends continue to grow forming curious pompoms. The splitting is thought to be caused by the creature's habit of digging the ground with its horns. It is coloured blackish-brown with white legs and a chestnut face pattern. The female is mainly grey in colour with lighter underparts and has upwardly curving, corkscrew horns, much like those of the African kudu. The colouration closely resembles the trees and foliage of the forest and, if standing still, the kouprey is almost invisible. According to those fortunate enough to have seen them, koupreys move with considerable grace, like antelope rather than cattle.

Kouprey normally inhabit low rolling hill country, with open grassland interrupted by thick forest. They like sandy soil and access to salt-licks. They feed mainly on the grass but need the forest stands in order to shelter from the hot sun, escape from predators, and find alternative food plants when the grasslands become parched.

Before the troubles began, a kouprey herd, led by an old female, numbered up to twenty animals, with several adult males and a bevy of females with their young. The herd did not always stay together but would split into smaller groups so as not to overgraze an area where food was in short supply. Sometimes kouprey were found in mixed herds with banteng or wild buffalo. There were also all-male herds and the occasional lone bull. Lone mature males were particularly fierce, and were sometimes seen butting tree stumps or termite hills.

Mating took place in April with young born in December. Calving followed the time when local villagers set fire to the dried-up straw cover and there were tender young shoots available to support suckling mothers. When a cow gave birth, it separated from the main body of the herd for four or five weeks.

The group of animals found in 1982 was in the Sisaket Province in northeast Thailand near the Kampuchean border, in an area dotted with land mines. One of the scientific party was injured when an attempt was made to capture and transport them to a captive breeding centre at Chockchai Ranch to the north of Bangkok. The group consisted of a bull, two cows and two calves.

The kouprey has been traditionally a respected beast. In Thailand, prehistoric cave-paintings depict the kouprey, and the ancient Khmers carved kouprey statues and featured it in bas-reliefs in temples, including Angkor Wat. Locally it is known as *ngua pho* or *ngua baa*. In 1964, Prince Sihanouk of Cambodia declared it to be the national animal, and indeed had a bull kouprey in the rose garden of the villa in

Phnom Penh where he was held under house arrest after having been deposed. The Prince fled, and the fate of the kouprey was unknown, the second only to have been held in captivity. The first was sent to Vincennes Zoo in Paris in April 1937, as part of a job-lot of animals from Saigon. It was not recognized and was wrongly labelled. It was eventually recorded for science in December of that year and became the third large mammal this century to have been discovered by the western world – the okapi in 1901 and the giant forest hog in 1904 were the other two.

Living Dragons

The Komodo dragon or ora is the largest known lizard in the world. It can grow to a length of 3m (10ft) and weigh over 165kg (364lb). It catches and eats animals as large as goats and pigs, and has even been known to attack and kill man. It lives on what were once remote Indonesian Islands – Komodo, where there are about 900 alive today, and in reserves on Flores, Rintja, Padar, Owadi Sami and Gili Mota, where there is estimated to be another 5,000. It is the only large carnivore on the islands and so, without competition, it has been able to evolve into an enormous animal.

The lizard was first recorded and described in 1912 by the curator of Java's Botanical Gardens at Buitenzorg. His 'type specimen' (the first actual specimen to be caught, killed, measured, weighed, dissected and published) was 2.9m long (9ft 6in.), and there have been others collected and measured at 2.66m (8ft 8in.) in 1915, 3m (9ft 10in.) in 1923, 2.76m (9ft 0½in.) in 1926, and 2.7m (9ft) in 1932. The 1926 specimen was just one of a staggering 54 taken by Douglas Burden on an expedition to Komodo.

The largest Komodo dragon known, and accurately measured, was a male specimen held captive at the St Louis Zoological Park in 1937. It was 3.1m (10ft 2in.) long and weighed 165.6kg (365lb). But there have been many more, probably exaggerated, claims. Two Dutch pearl fishermen informed the Governor of Flores in 1912 that they had killed several dragons with lengths up to 7m (23ft), and a Swedish zoologist in 1937 saw an individual he thought was about that length too.

The dragon's size is concentrated into a stocky body with a relatively short tail. It is not the longest lizard in the world; that honour must be awarded to the Salvadori monitor of the Eastern Highlands of Papua New Guinea, which has been reliably recorded on several occasions with lengths over 3.1m (10ft). Most of the length, though, is in the tail. Local villagers have killed lizards over 4.6m (15ft) long and claim they have seen animals over 6.1m (20ft). They call it *pukpuk bilong tri*, the tree crocodile. My colleague, Ian Redmond, whilst taking part in the 'Operation Drake' Expedition in 1980, saw a 3.7m (12ft) specimen that he was told had eaten a man two years previously. Salvadori's monitor has large, sharp claws but scavenges mainly on carrion, and is thought to avoid humans.

Komodo dragons ambush their prey, waiting in thickets of scrub until unsuspecting deer, wild pigs or feral goats, the legacy of ancient mariners, wander by. Despite their size and bulk they are surprisingly fast movers. They can also swim well. Their eyesight is poor and they rely to a great extent on smell to detect and locate their next meal. They frequently 'taste' the air with a yellow forked tongue that flicks continuously in and out of the mouth. They grab a goat, usually by a leg, and wrestle it to the ground where it is eviscerated. With powerful backward jerks of the head, large chunks of food are ripped from the carcass and swallowed whole. One 46kg (101lb) dragon is reported to have demolished a 41kg (90lb) pig at one sitting. The meal would have lasted the animal for several days.

Dragons have been known to follow a pregnant goat and grab the newly-born kid as it drops to the ground. They are also cannibalistic. Small dragons often supplement the diet of older giants. An injured person, or one suffering from heat exhaustion would be considered fair game for a dragon, although it is thought unlikely that they deliberately track humans.

A Fish's Revenge

The fishermen of Papua New Guinea have been chanting incantations to 'the forces of righteousness' in order that they may be saved from attack by an unusual creature – a fish the length of a man's foot.

This bizarre turn of events came to light when the new doctor arrived at the Provincial Hospital, Alotau, and was presented with a dead fisherman. Examination of the body revealed a small piece of sharp bone, resembling the end of a stiletto, sticking in the chest. A week later a second body was brought to the hospital and again it was that of a fisherman. This time the murderer revealed itself, for lodged, still alive and wriggling, in the victim's abdomen was a small, thin silvery fish – the needlefish.

Further investigations revealed that this fish had been responsible for up to twenty deaths a month. In the course of a single week, the doctor attended four fishermen who had succumbed to stab wounds in the chest or abdomen, three who were blinded in one eye, and two who were simply knocked unconscious.

All the dead and injured had one thing in common – they had been spear-fishing from a canoe at night using lanterns. The needlefish, possibly attracted by the light, had turned the tables on the fishermen and had leapt out of the water, spearing them with its 8cm (3in.) long, toothed snout.

A stab wound in the leg or arm is not considered serious, but clearly if the fish enters the chest or abdomen a fisherman could be killed, and so transfer to a hospital with surgical facilities is advised. The doctor suggested that canoes be placed in a circle, with a central pool of light, to lessen the risk of attack. One serious attack, though, on a three-year-old girl was in broad daylight.

Attacks from needlefish, whether intended or accidental, are not unknown from other parts of the world. An American fisherman, working a reef at night with a bright lamp off Miami, attracted the attention of a school of houndfish *Tylosurus crocodilus*, large relatives of the small needlefish which reach lengths of 1.4m (4ft 6in.). One enormous specimen shot out of the water and speared the fisherman through the leg, just below the knee, pinning him momentarily to the side of the wooden skiff. The fish thrashed about, tore itself free and leapt back into the water, leaving the man barely conscious because of the pain.

Needlefish, which are related not only to the garfish but also to the flying fish, have often been seen to catapult out of the water, breaking the surface at about 20 knots, and leaping 5m (16ft) into the air. More normal progression above the sea surface is attained by leaving the water at an angle, flying through the air like a javelin, and then, 20m (55ft) later, touching down with the tail. With a powerful flick of the tail, the fish propels itself off again and proceeds across the top of the water in a series of long hops. Alternatively, it can simply place its body at an angle of about 30° and skip along the surface by sculling with its tail.

The Swiftlet Caves

The caves of Malaysia, particularly those in the Gunong Mulu National Park of Sarawak, are considered to be the largest caves in the world. The cathedral-like galleries may be 122m (400ft) long and just as wide. Living in the darkness is an extraordinary collection of animals, many yet to be classified and recorded by science – crabs that behave like spiders, snakes that squawk and catch bats in mid-air, dinner-plate-size toads, blind amphipods and worms, shrews that squeak in the same frequency range as bats, and a small bird on which is based a multi-million-dollar industry for the production of bird's nest soup. The bird is the cave swiftlet and studying it has been Phil Chapman of the City of Bristol Museum.

The caves are in the remote rain forest and access is not easy. There is a two-hour drive along incredibly bumpy roads followed by another couple of hours trekking through waterlogged, leech-infested forest. 'You smell the caves before you see them', recalls Phil Chapman, 'the pungent smell of guano wafts through the trees.'

The entrances to the caves, green with the algae that line the walls, are enormous and have skylights in the roof through which shafts of sunlight penetrate. Further in, the light drops off rapidly and it is pitch dark. 'When your eyes become adjusted you see this strange fluted rock architecture where the water has run down the walls, and the whole lot is covered in a thin layer of guano. There are also huge piles of guano through which you must wade, and, in the places where water flows in, it turns into a quicksand which is really quite horrific.'

On one of his early visits to the caves he nearly came to an untimely and decidedly messy end: 'I didn't know about the bog-guano when I first visted one of the caves. Luckily I only stepped in the edge of it, but I was told that if I had gone forward instead of back, I would have probably disappeared without trace.'

The cave swiftlets are related to the common swifts that visit Europe in summer, but they are smaller and have a tubbier body and shorter wings. There are several species, some of which are able to echo-locate with clicking sounds and live in the inner galleries of the caves, and others that do not click and nest nearer the entrances.

The nest, which is built precariously on cave walls that overhang slightly, is started with a U-shaped strip of rubbery saliva, the 'hinge'. 'How they get a grip on the wall is not clear', says Chapman, 'but as soon as successive strips have been laid down then there is something on which to grip.' The nest is gradually built up, with thinner strips of saliva that are allowed to harden, into a cup-shaped shelf. Some of the nests consist of alternate layers of saliva and feathers, and appear striped. The nests of interest to Oriental gourmets are made only of saliva.

The nest collectors arrive in January and February when the first nests are built. To reach the nest sites, high above the cave floor, they build incredible scaffolding out of vines and pieces of bamboo wedged into cracks in the cave wall. The harvest is controlled. Only the first nests are allowed to be taken. Even so, many eggs and nestlings are turfed out on to the cave floor. The parents simply rebuild and attempt to raise another brood.

Some species of swiftlet lay two eggs, others only one. Once the youngster is hatched it remains in the nest for about forty days. When it is ready to fly it has a fundamental problem that the rest of the birds, other than oil-birds, do not have – it is in complete darkness. On its first excursion it must be able to fly well and have a workable echo-location system in order successfully to negotiate the passageways and galleries on its way to the cave entrance. Any error can be fatal, for sitting on the cave floor is a whole host of hungry creatures waiting for a meal.

There are snakes that are well able to navigate in the dark. They follow well-worn trackways in the guano – 'rippled ribbons of polished mud' – and are to be found 3–4km (2–2½ miles) into the caves where the swiftlets are nesting. They can climb the walls and sometimes take fledgelings from nests. They most likely detect their prey by smell, but one fascinating piece of behaviour observed by Phil Chapman suggests that it is not the whole story:

They also seem to catch swiftlets in flight. I actually saw one doing this. It twined itself around a stalactite, hanging down from the cave ceiling, and sat with its head out into the darkness and with its mouth open. The snake had chosen a place in the cave where the passage was very narrow and the swiftlets flew rather slowly, milling around in a kind of traffic-jam. In the confusion a swiftlet eventually flew into the snake, was grabbed and then eaten.

Chapman observed even more strange snake behaviour. He heard a snake squeak: 'They produce a hoarse mewing sound. I've heard this from only three or four feet or I wouldn't have believed it, because snakes aren't supposed to do that.' It could be that the snakes are using sound, as well as smell, to detect prey from a distance, although it is more likely that air pressure waves from the flapping wings give the snake the cue to strike in any particular direction.

The detection of pressure waves is common in cave animals. In the Malaysian caves there is a blind hunting spider, with surface relatives possessing good eyesight, that has a unique way of detecting prey. It runs on three pairs of legs and waves its front pair, which are covered with sensory hairs, ahead of it in the air. It is using the legs much like antennae to pick up air vibrations from the movements of small crickets, its usual prey.

Shrews, emitting very high frequency ultrasound – in the same frequencies, up to 100 kHz, used by the bats – are also lying in wait on the cave floor. They can detect injured or exhausted swiftlets and eat them. It could be that the shrews, well adapted for cave life with sensitive snouts covered by sensory bristles, use the ultrasound to investigate potential prey at a distance. If they nosed up to an animal such as a foraging cave snake that could bite back, it could be dangerous, if not fatal. The same species of shrew is commonly found in houses in Malaysia but these shrews have become dedicated cave-dwellers.

The most spectacular insect predator is a large cave cricket with great muscular legs and huge jaws. It is well able to tear a swiftlet apart, even in the nest. One has been seen to climb up to a nest, take an egg in its jaws, smash it against the rock, and eat the contents.

When the stream on the cave floor is in flood, crabs emerge to feed on debris washed out from the guano heaps and also on ditched swiftlets. They battle for the spoils, often ripping apart a chick between them. Catfish, admirably adapted for life in the dark with sensory barbles on the lower jaw, also live in the water feeding, not on the crabs, but on cave shrimps.

Further into the caves, where the creatures are even more adapted to a cave existence, there is another crab, related to the guano crabs, that has lost all pigment, has reduced eyes, and has very long legs. It scuttles around, both in and out of water, and is able to climb the walls much like an amphibious spider.

In the water are strange isopod and amphipod shrimps which appear to be related to species that only occur around the Mediterranean. Could it be that these species have been living in ground waters for so long that they were in existence before the continents started to drift apart? Might the inaccessible parts of the Malaysian caves be a time capsule that has trapped animals from a previous age? Phil Chapman thinks that this may be so. 'Once you've cracked the problems of living in darkness, there's no reason to change. It's a habitat that has existed for millions of years, so there are lots of living fossils that survive in caves.'

Many creatures visit the caves. There are dinner-plate-size toads. 'When you're walking through the caves', recalls Phil Chapman, 'you might go to stand on a rock and find that it hops away.'

Occasionally, sambar deer and bearded pig come in from the surrounding forest and seek out the guano heaps below the roost of free-tailed bats. Here they munch on the guano, using it much like a salt-lick, perhaps to make up some dietary deficiency.

Malay civets also visit the caves to scavenge on dead swiftlets and catch the larger crickets.

There are many species of bat living in the caves – fruit bats, which do not echo-locate and live near the cave entrances, mouse-eared bats, bent-winged bats, and the large diadem horseshoe bat that is the size of a fruit bat. The wrinkle-lip or free-tailed bat is fussy about where it roosts. Enormous colonies congregate only in the biggest caves with the largest entrances. They fly in large flocks and need space to be able to dodge the predators waiting outside. Bat hawks, dark peregrine-sized raptors (birds of prey) that hunt at twilight, swoop at an angle across the cave entrance, and tear through the emerging flock. Every morning and every evening there is pandemonium outside the caves as bats and swiftlets, both chased by the bat hawks, change shifts – the bats stream out and the swiftlets, diving in a spiral from a great height to avoid the hawks, fly in each evening; in the morning the swiftlets set out to collect insects while the bats return to the roost.

If it were not for the swiftlets and the bats this extraordinary and diverse cave community would not exist at all. They are the source of food. Both bats and swiftlets are collecting insects from a very wide area outside the caves and bringing it all back. In one cave alone as many as a million swiftlets might be living, so that the amount of food material brought in, and subsequently available as droppings, is incredible. 'There are far more creatures living on the floor of the caves', suggests Phil Chapman, 'than are living on the floor of the forest.' Many are species well known to science, but some, including many of the invertebrates, are as yet undescribed and awaiting detailed study.

13
The Oceans

The oceans cover over 70 per cent of the world's surface, yet they are the least explored of any of the major habitats. The oceans give us our weather, our water, but, curiously, contribute little to our supply of food.

The bottom of the sea has mountains, valleys, canyons, and plains much like those on the land. The Mid-Atlantic Ridge, for example, is a range of underwater volcanic mountains running, as its name suggests, in a north-south line in the middle of the Atlantic Ocean. The ridge marks the place where magma from deep down below the earth's crust is being pushed out to form the tectonic plates on which the continents sit. Where the plates collide, and one dives below the other, there are the ocean trenches, many far deeper than Everest is high.

The ocean currents can run in opposite directions at different depths. The surface currents are driven by the wind, while deeper currents are the result of differences in densities of water masses. Gigantic submarine waves race across the ocean floor.

Life is found throughout the sea, at all depths and under virtually all conditions – hot, cold, dark, light, high and low pressure, with or without oxygen, in sand or mud, under rocks, drifting free, on the surface or deep down.

At the Bottom of the Deep Sea

Before 1951, scientists speculated that life forms could not withstand the pressure and temperatures in the deepest parts of the sea, but in that year the Royal Danish Research Vessel *Galathea* was trawling on the bottom of the Philippines Trench, 10,000m (33,000ft) below the ocean surface, when it scooped up sea anemones, sea cucumbers and a bristle worm. Eight years later, Jacque Piccard took the bathyscape *Trieste* to the bottom of the Marianas Trench where he saw flatfish and some red shrimps. Baited cameras have revealed enormous, primitive sharks and small but prolific rat fish, deep-sea cod and brotulids scavenging on the dead animals falling from the waters above. Great knots of slimy hagfish, primitive jawless eel-like creatures, bore into the larger carcasses and eat them, from the inside out.

Undersea storms have been detected that have more energy and last longer than those in the atmosphere. Sediments are scoured from the bottom creating huge blizzards that settle and bury the animals living in the deep-sea muds.

Some of the creatures in the deep ocean glow in the inky abyss, their 'living lights' or bioluminescence the product of symbiotic luminescent bacteria living in special compartments in the creatures' bodies. Some fish use their lights for camouflage, others for communication or to lure prey.

Down to 4,000m (13,120ft), it is a world of 30cm (12in.) long monsters, such as the larva of the viper fish with eyes on elongated stalks, or the unique telescope-eyed giant tail which possesses binocular vision. Some of the most bizarre-shaped creatures live in the ocean depths – long, thin fishes with gigantic mouths and bellies that can expand to several times their normal size, their bodies mere appendages to sinister fang-filled jaws.

On the ocean floor itself, deep-sea sponges with intricate skeletons are anchored in the sediments by root-like structures. Deep-sea corals, with bulbous holdfasts, provide surfaces on which sea anemones and sea squirts can grow. Tall and slender sea lilies appear to grow out of the ooze like delicate rows of parking metres. In the mud, beard-worms up to 1.5m (5ft) long live in their chitinous 2.54mm (0.1in.) wide burrows, using thin tentacles to absorb food dissolved in the sea.

There are crabs that look like ordinary shore crabs and others with long spindly legs. Long-legged prawns and heavily-armoured shrimps search the sea-bed for scraps of food. A bright-red sea spider, with a minuscule body and eight very long and thin legs, measures 60cm (24in.) across. It pushes the end of its long proboscis into sea anemones and sea mats and sucks the juices. Deep-sea brittle stars hunt for prey, leaving criss-cross trails across the surface of the mud. Sea-cucumbers eat and sieve the mud for the organic particles on which they thrive.

Of the larger swimmers, there are many species of flat, bulbous-eyed deep-sea octopus and broad-finned squid. The tripod fish has long, thin extensions to its pelvic fin and tail with which its rests on the bottom, ready to spring into action after its prey. The male rat fishes and brotulids have modified swim bladders and associated muscles, and they can communicate with drumming sounds, evidence, perhaps, of courtship or territorial behaviour.

But, perhaps, the most remarkable discovery was made in 1977 when the manned submersible *Alvin* dived to the bottom of the sea off the Galapagos Islands in the Pacific Ocean and chanced upon an entire community of animals, never before seen by man, which lives in association with hot geysers on the ocean bed. It was the undersea find of the century, for the animals lived, not on food made with the energy from the sun, but with energy derived from the centre of the earth.

The community is able to thrive in such inhospitable conditions because hydrogen sulphide, in the hot water that seeps out through volcanic cracks in the sea floor, provides the basic food for marine bacteria. These, in turn, are food for 3m (10ft) long tube worms with bright-red plumes, giant brown mussells and 30cm (12in.) long white clams, spaghetti-like acorn-worms, amoeba-like protozoans with 12cm (5in.) long pseudopodia, bundles of yellow siphonophores resembling dandelion seed heads, strange pink 25cm (10in.) long deep-sea fish, and a multitude of bottom-dwelling crabs, shrimps with comb-like stalks where their eyes should be, and squat lobsters.

Since the initial discovery, hydrothermal vents and their unique animal communities have been found associated with mid-ocean ridges both in the Pacific and the Atlantic. Recently, hot springs have also been found near the Marianas Trench, in the western Pacific, where the Pacific plate is being forced down under the Philippine plate (a 'subduction zone'), and scientists are wondering whether it too will have its own unique animal community.

Furthermore, animals similar to those at hot springs have now been found to the west of Florida in the Gulf of Mexico. There is no volcanic activity, but it is thought that iron sulphide accumulates at the base of a range of undersea cliffs, known as the Florida Escarpment, having flowed down in dense, highly saline 'seeps' from the Florida Platform above.

In addition, researchers in California have found that animal communities associated with sulphur-eating bacteria are not confined to the deep sea. Off the Palos Verde Peninsula a diver found black abalone feeding on mats of white bacteria growing on miniature hydrothermal vents that seep from Los Angeles sewers. He also found a fish with the bacteria growing all over it 'like a white fuzz'. It seems that animal communities are likely to be thriving in all sorts of places, such as sewage outfalls and the areas around pulpmill waste pipes, that were formerly thought to be highly toxic.

Hot-Headed Fish

The swordfish *Xiphius gladius* has a heater in its brain. This fascinating discovery was made by Francis Carey, at the Woods Hole Oceanographic Institution, Massachusetts. He found that not only the brain, but also the retina in the eye and brown tissue on the underside of the brain case, can be up to 14°C (57°F) warmer than the surrounding seawater.

The temperature of a fish's body is dependent on the temperature of the surrounding seawater. Heat, produced by the activity of the fish's muscles, is retained in the blood until the blood vessels reach the thin-walled gills, where both heat and carbon dioxide are lost to the outside. Some fish have developed heat exchanger systems whereby organs that are required to work super-efficiently and instantly, such as swimming muscles, retain the body's heat. Tuna, mako, porebeagle and great white sharks do this, and are consequently some of the most powerful and fastest swimmers in the sea. If a fish is to remain alert and ready to respond to opportunities that might present themselves, it must make provision for warming up the organs involved, and this is just what the swordfish has done.

Examination of the swordfish head reveals that the mass of brown tissue, close to the brain and one of the eye muscles, is packed with parallel rows of small arteries and veins. Heat, about to be taken away by the veins, is transferred to the arteries and sent back to the brain. In addition, the brown tissue is filled with mitochondria (organelles which use up the oxygen to produce the principal energy-carrying compound in the cells of all living things), indicating a high level of metabolic activity and the production of more heat. Why, though, should the swordfish need such a system?

Little is known about the biology of the swordfish. It lives permanently in semi-darkness, swimming down to 600m (1,968ft) during the day and returning to the surface at night, a vertical migration in which the seawater temperature can change by as much as 19°C (34°F). It does not swim continuously, but is known as a 'stalker and sprinter'. Its muscles, unlike those of the tuna, are not in continual use, and so are at the same temperature as the surrounding water.

The swordfish, however, is an active predator that swims in short fast bursts in pursuit of swift-swimming prey. In order to spot and respond to passing food throughout the day, particularly at a depth where the water is cold, the animal warms up its eye muscles and its brain.

White marlin *Tetrapturus albidus*, and sailfish *Istiophorus platypterus* have the same system and can raise their brain temperature by 4°C (7°F). Blue marlin *Makaira nigricans*, striped marlin *Tetrapturus audax* and short-billed spearfish *Tetrapturus angustirostris* and a non-billfish, a large scombrid fish of the Southern Ocean *Gasteroschisma melampus*, also have the brown tissue but there have been no studies, as yet, to determine whether they have warm brains.

The function of the sword in the swordfish and the bills in billfish, incidentally, is not known. There are, however, many reports of the fish hitting boats, ships and whales and penetrating their timbers or blubber with the sword. Whether this is deliberate or accidental is not clear.

Swordfish can speed along at 100kph (60mph) and would probably find it very difficult to stop when chasing prey and suddenly confronted with the side of a ship or a whale!

Sex Change

For the first 46 days of its life, the Atlantic silverside *Menidia menidia*, common in estuaries along the Atlantic coast of North America, may change from male to female and vice versa, depending on the temperature of the water. In warm water, about 21°C (70°F), the fish becomes male and in cold water, around 15°C (59°F), it develops as a female.

Fish researchers David Conover and Boyd Kynard of the University of Massachusetts speculate that the ability to change sex in response to environmental influences has survival value. In the early spring, when water temperature is low, only females are produced and these, having a long spring and summer in which to feed up for the winter, are more likely to be alive the following spring to produce the next generation. Small males are hardier than females of the same age, so it is advantageous to produce males late in the season.

On the other side of the Atlantic, some members of the wrasse family go one better – they can be born one sex and later change into the other. The ballan wrasse *Labrus bergylta*, for example, relies totally on the ability to change sex to produce males. The fish spends the first fourteen or so years of its life as a female and only in 'middle-age' changes to male. Some females change earlier, round about five years old and not all fish make the transformation. There is usually one male to every ten females.

The male cuckoo wrasse *Labrus mixtus*, on the other hand, can be produced in one of two ways – some are born the same red colour as the females but are, in fact, males, known as primary males, while others start out as females and later change into blue-coloured males, and are known as secondary males. The primary males also change to the blue colour after several years, but they have small testes and appear not to take part in reproduction. Only red-coloured females and blue-coloured secondary males pair up. Francis Dipper, of the Nature Conservancy Council, who studied these fish off the Isle of Man, believes that the cuckoo wrasse is most likely on an evolutionary route to the ballan wrasse system, with the primary males being gradually phased out.

Breeding behaviour has not been observed in the ballan wrasse, but has in the cuckoo wrasse. The male excavates a nest hollow and attracts the female with the help of yet another colour change – in an instant his head loses all its colour. If the female is not interested, the colour immediately returns.

The male corkwing wrasse *Crenilabrus melops* also engages in an elaborate courtship display. Stimulated by the shape of the female, her abdomen distended with eggs, he dances in front of a nest constructed from fronds of seaweed. The female, if impressed, lays her eggs in the nest and the male guards them until they hatch, fighting off any other male, for instance, that comes near. Several females may be enticed to lay their eggs in the same nest. Some males, probably weaklings in the community, do not bother to attract females but become 'transvestites'. They simply mimic the female's body pattern and shape, fool a paired male into believing that they are females, and are allowed to approach the nest. When safely through the guarding male's defences, they deposit their own sperm and hopefully fertilize any unfertilized eggs that remain in the nest.

On Australia's Great Barrier Reef, the pygmy angelfish *Centropyge bicolour* has been studied by Janice Aldenhoven from Monash University. She found that these territorial fish live in harems with one male looking after and mating with about 5–10 females. If the male dies or is artificially removed, the largest female changes into a male. When the number of females exceeds 10, the largest female again changes into a male, takes some of the females, and starts another harem in new territory elsewhere. In areas where the fish are living in a high density, 'bachelor' males – females that have changed into males at an early age – may take over a harem when the harem-master dies, thus eliminating the need for a female to undergo a sex change.

Fish living in schools, such as a species of bass *Anthias squamipinnis* that lives in the Philippines and has been observed by Douglas Shapiro of the University of Puerto Rico, keep a constant ratio of males to females. In one group he found just two males with 13 females, and in the largest group 50 males amongst 294 females. When he removed 9 males from a group, 9 females changed sex and became males.

Most sex changes involve females becoming males, but clownfish, which live amongst the tentacles of sea anemones, change the other way. Hans and Simone Fricke, of the Max Planck Institute, have watched clownfish in the Red Sea and Indian Ocean, and have found that a typical family group consists of a large female, a smaller male and several juveniles. If the female is removed, the male changes sex and one of the juveniles becomes a fully functional mature male.

Sex change appears to occur in species of fish in which individuals of one sex are particularly successful in reproduction – when one large male, for instance, acquires, retains and mates with the breeding females, to the exclusion of smaller males. In this situation, it makes sense for a fish to be female when it is small, yet still in demand, and male when it grows bigger and can compete with the other large and dominant males. Sex change allows a fish to get the best of both worlds.

The Big Blob

About thirty miles northwest of Bermuda, a fisherman was bringing up a crab-filled fish trap that had been sitting on the bottom 800 fathoms below, when a strange thing happened:

> We were hauling up the trap and suddenly the winch began to run backwards. I jumped to the rope, which was a foolish thing to do as it was leaving the boat very quickly, and using water to stop the rope burning managed to stop it on the capstan roller. We started to haul again and as the trap reached about 200 fathoms we could see on the echosounder that there was something large on the trap. Then, there was a series of hard jerks, the line broke, and the trap was lost.

The fisherman and his partner were understandably shaken but continued to fish in the area. On several occasions they found something tugging on the line, a tug that sometimes amounted to a 2,700kg (6,000lb) pull. Their most eerie experience though was the last time they put down their traps to the northwest of the islands:

> I was trying to find ways of getting the trap back without damage so I snuck the rope up tight and did not try to force it off the bottom. It was quite a calm day with no swell and so I was able to manoeuvre the boat right above the trap. Then we switched off the engine and waited to see what would happen. Shortly after the boat began to move parallel to the shelf. It was towed for about a third of a mile. We had seen a large shape close to the trap and thought at first that it was a mass of crabs moving towards the bait. But, when the boat started to move I knew it couldn't be crabs.
> We were over a rocky bottom and could see this thing on the echosounder, moving from rock to rock. Then it released the trap.

The fishermen were fishing in very deep water in an attempt to find new fish stocks. They were working up the shelf slope from 1,900 fathoms and had found, at 500 fathoms, large numbers of cod, the national dish of Bermuda, and enormous numbers of very large deep-sea red crabs. One giant weighed 7.25kg (16lb) and was recognized as being similar to the golden crab of the Gulf of Mexico. Deep-sea crabs are taken by fishermen all along the Atlantic coast of North America. Their sweet meat is a fair substitute for the much-prized Alaskan king crabs. Also present in the traps was a new species of red crab identified as a relative of another new species found off the British coast. The fishermen believe that the crabs off Bermuda are associated with deep-sea hydrothermal vent activity, for their baits seem to deteriorate more rapidly than usual; an indication of warm water.

The main predator of the crabs is thought to be the conger eel, but speculation is rife that the crabs could be the principal food of a giant octopus.

There are, indeed, giants around Bermuda. Local fishermen some-times return with chunks of the giant squid *Architeuthis*, debris remaining after the violent encounters between squid and sperm whale. Giants squids were once part of maritime mythology themselves until specimens began to be washed ashore on the beaches of Newfoundland during the 1870s.

The squid differs from the octopus in having two long, grasping tentacles in addition to its eight shorter arms. A giant squid can weigh over 250kg (550lb) and measure 21m (70ft) from the end of the body to the tip of the outstretched tentacle. Horny parrot-like squid beaks recovered from the stomachs of sperm whales in Antarctic waters indicate that there might be even larger individuals lurking in the ocean depths. Not one of these giants has been seen intact, although judging by the numbers of beaks found, there must be a sizeable population in the Southern Ocean. A few years ago a US Navy submarine encountered a giant squid. The animal savaged the submarine's sonar array ripping the rubber housing. When the damage was examined a large hook was found embedded in the rubber. Several species of large squid have hooks inside the tentacle suckers, that are used to grasp prey. This hook, though, was three times the size of the largest previously recorded.

It is thought that giant squids travel in small groups. What they feed upon is unknown, although sports fishermen off the coast of Chile and Peru have witnessed the ferocity of their smaller 3.7m (12ft) long relatives, the Humboldt Current squids *Ommastrephes gigas*. A shoal of these animals can strip a huge swordfish or tuna in minutes, leaving only the head on the fishing hook. They are also cannibalistic. A squid caught on a line is immediately attacked by its companions. If a fisherman fell overboard accidentally, his fate would not be hard to imagine. These squids are considered amongst the most dangerous invertebrates alive.

There are only a few stories of giant squids harming man. The most horrific is the sinking of the troopship *Britannia* by a German raider in 1941. One night, with several survivors clinging to a small life-raft, a giant squid came up from the depths and wrapped its long tentacles around one of the men, dragging him below the surface. Another crewman had a tentacle around his leg but he was released and survived to tell the tale. To this day he is able to show the marks of circular weals on his leg – giant squid suckers over an inch across.

But what of a giant octopus? Is there any evidence that another giant cephalopod exists in the sea? The largest known octopus is the Pacific octopus *Octopus delfei* which inhabits the rocky Pacific coast of Oregon and Washington state. The record is held by an individual that was wrestled to the surface by a local skin diver in 1973. It had a radial span, from arm tip to arm tip of about 7m (23ft). Other specimens have been reported up to 11.6m (38ft) but the measurements have been unsubstantiated. They are considered to be relatively harmless and they avoid man by hiding in crevices among the rocks. Divers often catch them but there have been few cases of divers receiving bites from the powerful, horny beak at the centre of the eight radial arms.

In some parts of the world, octopuses have been known to attack and sometimes to kill. In September 1984 two fishermen from the island nation of Kiribati, who had been hunting octopus with spears, were said to have been held underwater and drowned by octopuses estimated to be 3.7m (12ft) across. In May 1985 the magazine *Diver* reported the frightening experience of a diver off Cyprus who had been taunting a large common octopus *Octopus vulgaris*. The animal took exception and enveloped the diver, tearing off his face-mask, and leaving him with a selection of purple and yellow circular bruises. In April 1935 a large Pacific octopus attacked a fisherman wading at the entrance of San Francisco Bay in waist-deep water. Another fisherman raced to the rescue, and, after a formidable struggle, the creature was despatched with a knife between the eyes. The octopus was reported to have had a radial span of 4.6m (15ft). In 1960 an oyster fisherman at Cape Agulhas, South Africa, was attacked by a large octopus which spared the man's life only to make off with his gold wristwatch.

The idea that the giant octopus might exist, not only in the imagination, but also in reality, is not new. Pliny told of a large 'polyps', with a head the size of a barrel and arms about 9m (30ft) in length, that would raid the fish ponds of Rocadillo in Spain. It is thought likely that Pliny's creature was an octopus and not a giant squid for only octopuses leave water and cross land to take their prey. Since those early days there have been many seafaring tales of octopuses so large that they could turn over a full-size sailing ship.

Modern science was not confronted with a likely specimen until November 1896 when two boys were on the seafront a little to the south of St Augustine, Florida. They chanced upon the carcass of an enormous sea creature. It had been washed ashore and lay partly buried in the sand.

Local GP and amateur naturalist Dr Dewitt Webb was summoned to the beach to examine the remains and in a letter published in the January 1897 edition of the *American Journal of Science* he described 'the body of an immense octopus . . . the body measures 18 feet in length by 10 feet in breadth.' In the same report the eminent cephalopod expert Professor Addison Emery Verrill, from Yale University, considered the animal to be a giant squid rather than an octopus. Later, though, he received pictures of the carcass and he changed his mind. In the April edition he declared that 'it is an eight armed cephalopod, and probably a true octopus of colossal size'. It was also reported that a piece of an arm, nearly 11m (36ft) long, had been found on the beach nearby. Calculations made at the time estimated that the creature had a weight of 7 tonnes and a radial span of 46m (150ft), thereby making it the largest invertebrate on earth. Later in the year Verrill withdrew his claims and proposed the carcass to be that of a decomposed whale, even though he had not actually seen it himself. Several whalers were sceptical about this interpretation for they could not understand which part of the whale the large sac-like remains could be.

There the story remained until 1957 when Forest Wood, then director of Marineland of Florida, was rummaging around his old files and discovered some yellowing press cuttings about the 'Florida Monster'. He was interested and began to look further into the event. He found that pieces of the creature had been taken to the Smithsonian Museum in Washington where they had been wrapped in cheese-cloth, preserved in formaldehyde, and kept in a glass jar labelled *Octopus giganteus*. By good fortune the samples were still intact and so Wood persuaded the Museum authorities to allow Joseph Gennaro, then at the University of Florida, to take some small samples for analysis. The tissue was tough. Gennaro dulled four sharp scalpel blades before he was able to cut away four finger-sized slices. His first observation was that the tissue was not oily like blubber. Neither did it have any distinguishing features on which Gennaro could base his assessment. Microscopic examination revealed that the pattern of the fibres was more akin to octopus than to squid or whale tissue. After a series of tests Gennaro concluded that the 'Florida Monster' was, indeed, an enormous octopus.

The scientific community, though, still refused to accept the identity given to the corpse. They required more proof. So, Gennaro enlisted the help of biochemist and cryptozoologist Roy Mackal from the University of Chicago. Mackal was sent samples of fresh squid and octopus, dolphin and white whale, preserved squid, and tissue from the 'Florida Monster'. He was unaware of the identity of each sample and was simply given a reference number. The object was to find out which amino acids were present in the samples in order to identify the proteins. Mackal found that one sample contained high concentrations of collagen, the main structural protein of animal connective tissue that has a high tensile strength but with little flexibility. It was later revealed that this sample came from the 'Florida Monster'. It is proposed that a large octopus would require just such a tissue to support its huge body.

Forest Wood, meanwhile, started to gather together anecdotal evidence that giant octopuses might be alive and well, and living not far from the Florida coast. In 1941 there was some concern that German U-boats might be hunting along the Atlantic seaboard of the USA and so the US Navy carried out a major depth-charging programme that, as one person put it, 'rearranged the ecology of the entire ocean floor'. A lookout stationed on a vessel following the action reported seeing a large brown kelp-like mass floating at the surface, not far from Fort Lauderdale. As the ship moved closer the sailor saw that it was not seaweed. 'As it moved into view there was no doubt as to its identity. The coils of its arms were looped like huge coils of manilla rope.'

Wood also examined the movements of the ocean currents and worked out the likely area from which the 'Florida Monster' might have drifted. The target turned out to be the Bahamas, particularly deep water channels around Andros Island. Wood visited the islands and discovered that there were many stories told by the local fishermen of giant octopuses. The Bahamians call them giant scuttles or lusca. When one fisherman was asked to indicate their size he simply pointed to a shed some 23m (75ft) away. Local folklore held that the creatures entered shallow water only when sick or dying, and that they were only dangerous if they were firmly attached to both the bottom of the sea and the bottom of the boat at the same time.

In one story an island official recalled an encounter with a giant scuttle during a fishing trip with his father. They were off Andros Island in about 180m (600ft) of water when his father thought he had snagged the bottom. The line, though, could be slowly drawn up. As the hook and bait came nearer to the surface, father and son peered over the side and through the crystal-clear water they could make out the shape of a gigantic octopus. The animal suddenly detached itself from the line and clung to the bottom of the boat. Much to the occupants' relief, it then let go and sank back into the depths.

So, might the creature playing tug-of-war northeast of Bermuda be a relative of the Bahamian lusca? There are, it seems, plenty of curious cephalopods known to be living in the area. Sean Ingham caught another giant though not, he points out, the same creature that is living on the bottom. He was hauling up a crab-trap from about 500 fathoms and felt a strong pull on the line. As the trap approached the surface, he could see what appeared to be a very large cephalopod with large 'tentacles' wrapped around it. He tried to hook it but the gaff just pulled through the tissues and pieces broke off. As the trap was pulled clear of the water the entire animal began to break up on the wire-mesh. In the end, he was only able to recover a small piece in a bucket. The chunk was photographed and then placed in the freezer for safe keeping. The pictures were sent to the Smithsonian but contain insufficient detail for an identification. The jelly-like appearance of the creature suggests any number of marine animals. Giant 2m (6ft) long colonial polyps, such as *Pyrosoma*, have been found drifting in the ocean, but Sean Ingham believed it was an octopus with a 9m (30ft) radial span and not a polyp that he had brought up. One contender looks more like a jelly fish than an octopus and that is the deep-sea octopus *Aloposis*. It resembles a large gelatinous umbrella, with short arms and big eyes.

The gelatinous octopuses were once thought to be rare, for very few have been caught in research nets. When the USA lost its nuclear warhead off Spain, however, the numerous photographs taken of the sea floor revealed umbrellas all over the place. They are known to grow as large as 2m (6ft) across – 1.5m (5ft) of body and the rest in arms, but it is thought that there may be even larger specimens living in the deep. The small varieties, such as *Vitreledonella*, are found in mid-waters, while the ones with fins, such as *Cirroteuthis*, live on the bottom. *Aloposis* is usually found above shelf slopes: just the sort of place through which Sean Ingham's traps must have passed on the way back to the surface.

Recently, the Marine Biological Association laboratory at Plymouth was sent photographs of a very large gelatinous octopus that had been brought up in a fishing net from 100 fathoms in the Atlantic, west of Ireland. Its most striking feature is the size of its eyes: they are 15cm (6in.) across. The identity of the creature remains a mystery.

14

A Mystery from the Living World

In the reign of King Nebuchadnezzar about 600 BC, an unnamed Babylonian artist shaped bas-reliefs of several strange animals as part of the maintenance programme on the structures associated with the Ishtar Gate. Over the centuries his magnificent work was buried and it was not until 3 June 1887 that German archaeologist Robert Koldeway stumbled upon a brick with one face coated by a blue glaze. The professor had rediscovered the Gate, which he later excavated in 1902.

The enormous semicircular arch and the high walls of the approach road are faced with bricks, each glazed blue, yellow, white and black. The animals are in alternating rows with lions that look like lions, fierce bulls that do not look quite like everyday cattle, and curious long-necked dragons with a serpent's head, forefeet like those of a big cat, but hind feet more in keeping with those of a bird, talons and all. The Babylonian names were preserved in cuneiform inscription – the lion is identified by the common term for lion, but that for the bull is a special one, *rimi*, and the dragon is *mushhushshu* (from the Sumerian: *mush* meaning snake and *hush* meaning frightful).

The fierce-looking bovine could be identified as the aurochs – a wild ancestor of our domestic cattle and a creature that was extinct in Mesopotamia at the time the Gate was constructed but which lived in Eurasia until the final one disappeared from Lithuania in 1627. On what, though, had the Babylonian artists modelled their dragon? Was it simply the figment of an unbridled imagination, or perhaps based on traditional animals in mythology – the *mushhushshu* is after all thought to be one of the Babylonian constellations – or did it have some relationship, as the other sculptures had done, with a real living creature?

The *mushhushshu* had been mentioned elsewhere – in the Book of Bel and the Dragon from the Apocrypha. A strange creature was kept in the temple by the priests of the supreme god Bel, Lord of the World, and the people of Nebuchadnezzar were encouraged to worship it, that is until Daniel came along:

And in that same place there was a great dragon, which they of Babylon worshipped. And the king said unto Daniel, Wilt thou also say that this is of brass? lo, he liveth, and eateth and drinketh; thou

canst not say that he is no living god: therefore worship him. Then
said Daniel, I will worship the Lord my God: for he is a living God.
But give me leave, O king, and I shall slay this dragon without sword
or staff.

And that is just what Daniel did. He forced lumps of bitumen, laced with
hair and fat, down the dragon's throat and burst it apart.

The *mushhushshu* also featured in the paintings and seals of the god
Marduck which had, as its personal symbol, a horned dragon. These
pre-dated the reliefs on the Gate by over a thousand years.

But had the Babylonians really kept a living dragon? Could it have
been an exceptionally large snake or monitor lizard? Koldeway
proposed a more bizarre identity. His excavations were at a time when
dinosaurs had been described and named and their fossil bones had been
causing considerable interest in the scientific community, so, he stuck
his neck out and likened the Babylonian dragons of the Ishtar Gate to
extinct dinosaurs. Indeed, he went on to place them with the bird-hip
dinosaurs and in 1918 wrote, 'The iguanadon of the Cretaceous layers of
Belgium is the closest relative of the Dragon of Babylon.'

There were, at that time though, no fossils of dinosaurs available from
that region and it is unlikely that Babylonian artists would have
reconstructed fossilized animals anyway. Could the dragon then, like
the aurochs, be known from distant lands? They would not have had to
travel far to meet the *afa*, a river monster with legs, and known to the
Marsh Arabs who live in the wetlands of the Tigris River.

The Babylonians were well-travelled people, and although there is no
firm evidence of them penetrating equatorial Africa, there is one
intriguing piece of anecdotal evidence that they had some link with
West and Central Africa – just the sort of secluded place where a relict
species might persist. According to accounts in Willy Ley's *The Lungfish
and the Unicorn*, Hans Schomburgh, a big-game hunter from Germany,
allegedly found a glazed brick in the African rain forest, significantly
some time before Koldeway had found the Gate, and he took it back to
his boss, the director of Hamburg Zoo, Carl Hagenback. Did this mean
that the Babylonians might have been in this area hundreds of years
before and had encountered a strange and powerful creature that had
caught the imagination and was worshipped as a god? It seems unlikely –
after all, if glazed bricks were found in West Africa, they are more
likely to have reached there as part of the ballast of sixteenth-century
and seventeenth-century sailing ships. Babylonian stone monuments,
found in the foundations of a seventeenth-century house in the City of
London, were thought to have arrived on English shores in that way.

Schomburgh was in search of a creature known to the locals as the *nygbve*, now known as the pygmy hippopotamus. He returned to Europe, not only with both living and dead specimens of *nygbves*, but also incredible tales of curious creatures living in the forest. There was a pygmy rhinoceros in the mountains of eastern Liberia, and a vicious animal, as large as a goat but with teeth like a dog, called *too*. Then there was the giant upright walking primate, with long black hair and a white face, known to the pygmies as *muhalu*, and the 'river elephant' from which Schomburgh saw a sample of skin covered in long red hair. But the most interesting tale of all was that of a 'huge monster, half elephant, half dragon' that was reputed to live in the swamplands and chase away the hippos. On reaching Lake Bangweolo, Schomburgh noticed that, indeed, hippos were absent although conditions for them were ideal. The local inhabitants blamed the 'dragon'.

Hagenback, intrigued by Schomburgh's stories, sent another expedition to Lake Bangweolo. In his book, published in 1912, he recalls, 'On the walls of caverns in Central Africa there are actual drawings of this strange creature. From what I have heard of the animal, it seems to me that it can only be some kind of dinosaur, seemingly akin to the brontosaurus. As the stories come from so many different sources and all tend to substantiate each other, I am almost convinced that some reptile must still be in existence. At great expense, therefore, I sent out an expedition to find the monster . . .'

Unfortunately malaria got the better of them and not only did they fail to find the creature, but also they could not find the lake.

Many early accounts of the fauna and flora of West and Central Africa came from missionaries and merchants. In 1776, the Abbe Lievain Bonaventure Proyart wrote, in the *History of Loango, Kakonga, and other Kingdoms in Africa*, about a group of missionaries who had found the tracks of an unknown animal in the forest. Pinkerton's translation, published in 1814, reads:

> it must be monstrous, the prints of its claws are seen on the earth, and formed an impression on it of about three feet in circumference. In observing the posture and disposition of the footsteps, they concluded that it did not run in this part of its way, and that it carried its claws at the distance of seven or eight feet one from the other.

It has been estimated that such marks could only be made by a creature the size of an elephant, and there is, indeed, a subspecies of elephant – the forest elephant – living in the region. It differs from its bush-living relative in having round ears, straighter downward-pointing tusks, finely wrinkled skin, and a head that is carried low. It does not, however, have large claws.

Alfred Aloysius Smith, otherwise known as 'Trader Horn', whilst plying the Ogooue River in the Gabon during the late nineteenth century, was told stories about an enormous dragon-like river beast called *jago-nini*. And while travelling later in the Cameroons he visited lakes from which populations of manatees were said to have been wiped out by the *amali*, a creature that left three-clawed frying-pan-sized footprints in the mud at the side of the lake.

More reliable scientific reporting did not emerge until the year after Hagenback's unsuccessful adventures when in 1913 the German Government decided that its colony, the Cameroons, should be surveyed. Captain Freiherr von Stein zu Lausnitz was the chosen official to lead the expedition and in his report he devoted a substantial section to Schomburgh's mystery animal, in particular to the references in the 'narratives of natives'. The Captain was understandably cautious but drew attention to the fact that his informants were 'experienced guides who repeated characteristic features of the story without knowing each other'. They told of a 'creature feared very much by the Negroes of certain parts of the territory of the Congo, the lower Ubangi, the Sangha, and the Ikelemba rivers'. They called the animal *mokele-mbembe*. Captain von Stein wrote:

> The animal is said to be of a brownish-grey colour with a smooth skin, its size approximating to that of an elephant. It is said to have a long and very flexible neck and only one tooth but a very long one; some say it is a horn. A few spoke about a long muscular tail like that of an alligator. Canoes coming near it are said to be doomed; the animal is said to attack the vessels at once and to kill the crews but without eating the bodies. The creature is said to live in the caves that have been washed out by the river in the clay of its shores at sharp bends. It is said to climb the shore even at daytime in search of food; its diet is said to be entirely vegetable.

In 1920, it was reported in *The Times* (although, more recently, it has been questioned as a hoax) that a railway construction engineer named Leplage was out hunting and was scared out of his wits by 'an extraordinary monster, which charged at him'. Leplage fired at the beast but was forced to retreat. The creature gave up its chase and Leplage was able to watch it through binoculars. The article continued: 'The animal, he says, was about 24ft in length with a long pointed snout adorned with tusks like horns and a short horn above the nostrils. The front feet were like those of a horse and the hind feet were cloven. There was a scaly hump on the monster's shoulder.'

Monster sightings were legion during the early part of the twentieth century and many of them are documented in Bernard Heuvelmans' monumental collection of tales of strange creatures, *On the Track of Unknown Animals*. They were not confined to the Congo. A long-necked 'prehistoric' beast was described by officers aboard the Lake Victoria steamers, and Deputy British Governer H.C. Jackson, who published a study in 1923 on the Nuer people of the Upper Nile Province, wrote about the *lau* of the swamplands at the source of the White Nile. The King of Barotse saw a huge animal in his marshes alongside the Zambesi. He was so impressed that he sent an official report to the British Government mentioning that the creature had a head like a snake, and that it made a huge track in the reeds as large as a fullsized wagon would make if it had no wheels.

In the Dilola swamps of Angola an intrepid South African hunter by the name of Grobler was told stories about the *chepekwe*, a four-ton monster with the head and tail of a lizard, that attacked rhinos, elephants and hippos – 'crushing the bones and tearing out huge lumps of meat'. Another fearless South African plunged into the rainforest near Lake Edward, on the Ugandan–Zaire border, in pursuit of the *irizima*, a monster with a hippo's legs, an elephant's trunk and a lizard's head. He claimed he had seen a brontosaurus tearing its way through the reeds, but unfortunately could not manage to follow it.

In 1938, Dr Leo von Boxenberger, an ex-colonial magistrate, explored the central and southern parts of the Cameroons where reports of *mokele-mbembe* could be collected, but he lost all his notes in an attack by the Pangwe tribe in Spanish Guinea, and subsequently little was heard of the dragon-like dinosaur monsters of Africa.

Then in 1976 crocodile expert James Powell was working along the Ogooue and N'Gounie rivers in Gabon when he heard about the *n'yamala*. The stories matched those of Trader Horn's *amali*. Powell conveyed the information to Roy Mackal, the biochemist (mentioned earlier) at the University of Chicago and Vice-president of the International Society of Cryptozoology, an organization of eminent scientists that serves as a focal point for the investigation of matters related to animals of unexpected form or size, or unexpected occurrence in time or space. Mackal became hooked by the history, in particular the stories of *mokele-mbembe*, and agreed with Powell that an expedition should be mounted. First they had to choose the best area for a search.

Mackal felt that Gabon was not the centre of *mokele-mbembe* activity. The stories were told by the Fang people, a Bantu group that had probably travelled to the coast from further east. Powell's contacts had probably been retelling legends passed down through several generations. By tracing back their likely migration route he established a target area between the Sangha and Ubangi rivers, a region of impenetrable swamp forest left blank on most maps.

In 1979, Mackal and Powell set out from Brazzaville to reach Impfondo where they met Eugene Thomas, a missionary who had worked and lived in the Congo for 25 years. Thomas had heard many stories about *mokele-mbembe*, and while the expedition was at the mission an eyewitness turned up. He told of having seen the creature on a bend in the Likouala aux Herbes river near Epena. Unfortunately Mackal and Powell had missed the local light aircraft flight and because their time was limited they decided to set off on foot across the vazierre – the great swamp. They trudged through thick mud that released a cloud of stinking methane and hydrogen sulphide at every step. The temperature was a constant 32°C (90°F) and the humidity high. They were under constant assault from mosquitoes, ants and a multitude of other biting and sucking insects, and were stalked by highly poisonous snakes, such as like river jacks and green mambas.

Having spent so long travelling, there was little time left for investigation, but nevertheless they gathered more stories and found the plant on which *mokele-mbembe* is reputed to feed. Local pygmies call it *molombo*. Botanists call it *Landolphia*, a liana with white flowers, a sweet white latex sap, and fruit similar to apples. *Mokele-mbembe* likes the apples and emerges in the late afternoon to feed, a time when the pygmies avoid being on the Likouala aux Herbes river.

Mackal recalls the stories coming thick and fast:

> The pygmies described to me animals that were 15–20 feet long, but most of that length is head, neck and tail. They told me that the largest had a head and neck like a snake the size of my thigh. They said that the tail was very long and thin, but that the body was more bulbous like that of a hippo. The largest ones were the size of a small elephant. We were told that they had stubby legs, and that the hind feet each have three claws. The animals are reddish-brown, have no hair, but have a comb-like frill running down the back of the head and neck, a description which tallies with reports for the past 200 years. They said that they live in rivers, lakes, and streams, and that they are very rare and dangerous. Although they eat plants, they can upset canoes and kill the occupants. The pygmies indicated that they are very much afraid of the animals.

One of the most bizarre stories that Mackal and Powell took down described the events that were reputed to have taken place in 1959 at Lake Telle. Mackal recollects:

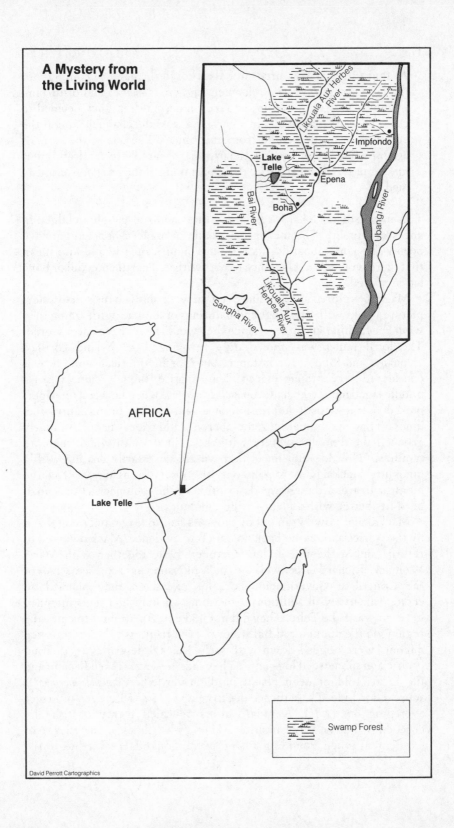

A Mystery from
the Living World

Likouala Aux Herbes River

Impfondo

**Lake
Telle**

Epena

Boha

Bai River

Ubangi River

Likouala Aux
Herbes River

Sangha River

AFRICA

Lake Telle

Swamp Forest

At that time, two or three animals were in the habit of coming into the lake along a series of channels known as *molibos* and disturbing pygmy fishing activities in the lake. In desperation, they decided to cut stakes and place them in a barrier across the channels so that the creatures were prevented from entering. When one of the animals attempted to break through, the pygmies speared it to death and cut it up. Some ate the meat but all those partaking of the feast mysteriously died.

This interested Mackal for he knew that the local people indulged in ancestor worship. This meant that a skull, or maybe a bone, might have been kept. Unfortunately, its religious significance would also mean that there would be a reluctance to part with it. To date, no skull or bone has appeared.

Mackal explored another line of enquiry: he showed the story-tellers picture books with drawings and paintings of animals with which they would be familiar and others which they would never have encountered. The local fauna was immediately recognized – lowland gorillas, elephants and hippos, for instance, but North American wildlife was considered very strange indeed. The surprise, though, came when a middle-aged man, who had seen *mokele-mbembe* when he was a teenager, picked out an artist's impression of a sauropod (a brontosaurus-like dinosaur) as the animal he remembered. Other local people similarly grouped the dinosaur picture with the animals with which they are familiar. 'This does not mean that we are necessarily dealing with a dinosaur,' Mackal is fast to point out, 'the observations may be of a large monitor lizard, although the description of the long neck with a small head fits better with a plant-eating dinosaur'.

Mackal and Powell ran out of time and had to leave, only tantalized by the stories and taking back no concrete evidence. Mackal vowed to return, and so he did in late October 1981, together with Maria Womack, Richard Greenwell, Justin Wilkinson, and representatives of the Congolese Government. For this excursion they planned an ecological survey in addition to the dinosaur hunt and their intended location was Lake Telle. They did not make it. An attempt to enter the region via the unexplored Bai River proved fruitless. Their motorized dugouts were bogged down with what Mackal describes as 'floating prairie and submerged log-jams'. They did, however, reach Kinami and the locals told of deep places in the river where *mokele-mbembe* is supposed to hide. These turned out to be up to 15m (50ft) deep in a river averaging 6m (20ft). A sonar survey revealed plenty of fish and crocodiles, but no large monster. There was, though, one flicker of excitement when, rounding a bend in the Loukouala river just to the north of Epena, the expedition members spotted a 12cm (5in.) high wake that suggested the submergence of a large creature. Crocodiles do not make such a wake and hippos are not present in the area. But they saw no animal . . .

On another occasion, a hunter took them to see a trail that had been left by a large animal. He said that he had first thought that it had been made by an elephant but there was something curious about it. It led into the water but did not leave on the other side in the way that an elephant is more likely to go.

Time beat the expedition again, but before they left they were able to make some important observations about the flora and fauna of the region. They noted, for example, that the numbers of many of the large animals were diminishing. The population of forest elephants is fast becoming a statistic included in the world trade in ivory. There are no controls. Gorillas and chimpanzees are killed whenever encountered and monkey colonies close to villages have been decimated. Monkey flesh is an important protein source for the local inhabitants. So too are the various species of fish including the African lungfish *Protopterus*, the characin known locally as the leopard-fish, and a mud skipper that dangles its tail in the water to obtain oxygen.

At about the same time, another US-Congolese expedition led by Herman Regusters, an engineer on leave from the Jet Propulsion Laboratory in Pasadena, did make it to Lake Telle: he saved valuable days by flying directly into Epena and travelling by dugout to the village of Boha. It is the Boha villagers who consider themselves to be the rightful 'owners' of the lake and without their cooperation it is impossible to enter the region. Here, like Mackal, Regusters found that all the maps were wrong and that river access to the lake is impossible. There is, however, a tongue of land that become known as the 'Boha ramp' which enabled the expedition to wade to the lake, a damp trek which took five days. Regusters remembers conditions as being tough:

Within the forest the great broadleafed trees form a dense canopy overhead, allowing only a slight penetration of sunlight or rain. Giant woody vines or lianas are suspended from tall trees, and dense thickets of shrubs and smaller vines create an almost impenetrable mass . . . For the five days of our trek we remained completely wet. In the evenings, our small tents were erected on beds of leaves that had accumulated over the years on the great aerial roots of the tall trees . . . In the forest we were always besieged by swarms of sweat bees whenever we halted to camp . . . Great colonies of ants and termites caused damage to nylon tents and netting and to leather goods.

On the last day of the 'ramp', the expedition was forced to walk through red-dyed water, sometimes chest high, in order to reach the lake. Again Regusters found the maps to be wrong. The channels on the western side of the lake turned out to be long lobes of the lake that merged into the jungle. The native guides were fearful, particularly, of lobes 6 and 10. They were afraid, too, of the centre, expecting their boats to be sucked down to the bottom by some mysterious force. The expedition measured strong currents towards the centre of the lake.

One evening, expectation rose to fever pitch when a fast-moving disturbance was seen on the glass-smooth water. The following morning a dark 'long-necked member' was spotted and kept under observation for about five minutes. It was seen again on two occasions, a few days later. On an excursion to the southern end of the lake, Regusters and his wife Kai heard an extraordinary sound, 'starting with a low windy roar, then increasing to a deep-throated trumpeting growl', which accompanied the movement of a huge creature through the swamp at the edge of the lake. A couple of weeks later a large 'dark-brown object' was seen to leave a wake at the surface close to the infamous lobe 10. All the expedition members agreed that it was not a hippopotamus, for they do not live in the region, nor a crocodile, which would not show such a profile above the water.

A couple of days later, again near lobe 10, a 5m (16ft) long neck with a small head emerged from the water. It was about 30m (96ft) from the expedition's boat. The head and neck wavered from side to side and then sank vertically down below the surface. The expedition unfortunately were unable to bring back any photographic evidence – all their cameras had failed because of the high humidity – but they did bring back another curious tale. In February 1981, three adult elephants had been found dead, floating in the water with two punctures in their abdomens. Their tusks were still intact, eliminating poachers as a likely cause of death. The local villagers thought responsibility lay with a rhinoceros-like beast that lives in the forest. Could there be another unknown giant living in the forest?

In April 1983, Marcellin Agnagna, from the Zoological Park of Brazzaville, led a Congolese expedition into the Likouala region in search of *mokele-mbembe*. They reached Boha and, after some reluctance on the part of the villagers, were eventually allowed access to the lake. Drums were beaten to 'evoke the spirits of their ancestors, to protect the expedition members, and ensure the success of the mission to Telle'. It was the dry season and the water level was low, so the trek took just two days. On reaching the 5km (3 miles) by 4km (2.5 miles) oval lake they watched an enormous freshwater turtle which had, it was estimated, a shell 2m (6ft 6ins) long. These large reptiles were numerous in the lake, providing food for the expedition members and war shields for the villagers.

On 1 May, a cry from one of the local guides brought Agnagna and his colleagues to the water's edge and there they witnessed 'a strange animal, with a wide back, a long neck, and a small head. The frontal part of the animal was brown, while the back part of the neck appeared black and shone in the sunlight.' The movie camera, in the heat of the moment – surprise, surprise – failed to get the definitive pictures. The macroswitch was in the wrong position and the image is just an out-of-focus blur. In a scientific paper Agnagna wrote: 'It can be said with certainty that the animal we saw was *mokele-mbembe*, that it was quite alive and, furthermore, that it is known to many inhabitants of the Likouala region. Its total length from head to back visible above the waterline was estimated to be five metres.'

The scientific community is not convinced by the reports and so yet more expeditions are needed to get the definitive pictures or an actual specimen. At the end of 1985 and beginning of 1986, a well-equipped British-Congolese expedition fought its way through the swamp forest but failed to find any monsters (although they believe that they discovered a new species of monkey).

Roy Mackal, together with other members of the International Society of Cryptozoology, is planning his third attempt to find the elusive beast in 1988, but what is he really likely to find? Could *mokele-mbembe* be a dinosaur that somehow survived the 'time of great dying' at the end of the Cretaceous about 64 million years ago? Roy Mackal does not rule out the possibility:

We think that small sauropods might have survived in this area. It has not changed appreciably since the Cretaceous. The vegetation is very primitive. The *molombo*, for instance, belongs to the Apocyanaceae and was in its heyday in the Cretaceous. There have been no major climatic changes. The temperatures are still the same. There have been no mountain building episodes here. The environment of the Congo basin has remained relatively stable for all those millions of years. So, if there is a relict species, it is perfectly reasonable that it may remain undetected in an area of remote unexplored swamp-jungle that is malaria ridden, full of poisonous snakes, and is a most unpleasant and formidable place to be.

There is a precedent. At the beginning of this century, Sir Harry Johnson went in search of a mysterious creature that had been described by the Mbuti pygmies of the Ituri Forest as being something between a giraffe and a zebra. 'Balderdash,' cried the scientific establishment. 'It's a mutant zebra.' A creature fitting that description has been found as a fossil. It is called *Palaeotragus*, and when it had lived it was a moderate-sized browser with two short bony horns. It was considered to have become extinct about 20–30 million years ago. In 1900, Sir Harry found an animal alive and well that looked just like *Palaeotragus* and fitted exactly the stories told by the pygmies. We call it the okapi.

Bibliography

The Encyclopedia of Mammals, Vols 1 & 2
Edited by David Macdonald
Published by Unwin Animal Library 1984

Authoritative and accessible accounts of the biology of mammals worldwide. The best two volumes from a series of very readable encyclopedias.

A Dictionary of Birds
Edited by Bruce Campbell and Elizabeth Lack
Published by T. & A.D. Poyser 1985

Excellent reference work and a must for the library shelf of anybody interested in the life of birds.

Animal Language
by Michael Bright
Published by BBC Publications 1984

My own account of the way that animals make use of sound, particularly for communication.

The Oxford Dictionary of Natural History
Edited by Michael Allaby
Published by Oxford University Press 1986

A very useful and comprehensive look-it-up book.

The Book of Sharks
Written and illustrated by Richard Ellis
Published by Robert Hale 1983

A beautifully-illustrated book with well-researched text gives you all-you-need-to-know about sharks.

The Times Atlas of the Oceans
Edited by Alastair Couper
Published by Times Books 1983

A unique account of the physical and biological aspects of the oceans, with detailed information about man's uses and abuses.

i *On the Track of Unknown Animals*
ii *In the Wake of Sea-Serpents*
by Bernard Heuvelmans
Translated from the French by Richard Garnett
Published by Rupert Hart-Davis (i.1962 and ii.1968)

If you can get hold of copies, these are still the authoritative works on terrestrial 'unknowns' and sea serpents from the President of the International Society of Cryptozoology.

Cetacean Behaviour
Edited by Louis Herman
Published by Wiley-Interscience 1980

An academic work but full of the latest research on whale and dolphin behaviour, including papers from the leading cetacean researchers in the world.

Collins Guide to the Insects of Britain and Western Europe
By Michael Chinery
Published by Collins 1986

This is one of the best general insect guides so far produced with excellent illustrations, informative text and a wide-ranging selection of insects and other terrestrial invertebrates.

Gorillas in the Mist
By Dian Fossey
Published by Hodder and Stoughton 1983

The story of the late Dr Fossey's work in Rwanda with the very rare mountain gorillas.

BBC Wildlife Magazine
Edited by Roz Kidman Cox
Published by BBC Publications

This beautifully illustrated monthly magazine brings you all the latest news and views about natural history, behaviour, ecology, conservation and a whole lot more – a more permanent record of those stories you might first hear about in 'The Living World'.

Index